Springer Hydrogeology

More information about this series at http://www.springer.com/series/10174

Constantin Moraru · Robyn Hannigan

Analysis
of Hydrogeochemical
Vulnerability

 Springer

Constantin Moraru
Laboratory of Hydrogeology
Institute of Geology and Seismology
Chisinau
Moldova

Robyn Hannigan
School for the Environment
University of Massachusetts Boston
Boston, MA
USA

ISSN 2364-6454   ISSN 2364-6462   (electronic)
Springer Hydrogeology
ISBN 978-3-319-89021-0   ISBN 978-3-319-70960-4   (eBook)
https://doi.org/10.1007/978-3-319-70960-4

Printed on acid-free paper

This Springer imprint is published by Springer Nature
The registered company is Springer International Publishing AG
The registered company address is: Gewerbestrasse 11, 6330 Cham, Switzerland

*This book is dedicated to our children
Camellia Hannigan—Cristian,
Darya Moraru,
Sophia Moraru—Krijanovskaia
may be as future geochemists*

Preface

Groundwater is used for many purposes: drinking, irrigation, industrial processes, and as valuable sources for rare chemical elements like lithium, uranium, helium, and others. All groundwater is stored in aquifers. Despite of underground position, aquifers need to be protected. Ecological groundwater protection is more complicated compared with surface water. Groundwater is everywhere, but at different depth and in different geological conditions. In the past 10–15 years, the enormous impact of human activity, exemplified by point and non-point sources of pollution, has been recognized. It is well known that future protection of groundwater must be based on the understanding of the role of natural unsaturated zone that regulates the transport and delay of pollutants. This is the basic principle of the groundwater vulnerability assessment.

It has been stated that the geological information, available to specialists, is now doubling every 10 years. The purpose of this book is to inform reader of our present knowledge about the hydrogeochemical vulnerability. We introduce notions of geochemical signal, quasi-point of migration of pollutants in the unsaturated zone and new hydrogeochemical classification based on compositional data, and our view related to basic hydrogeological elements is needed for vulnerability research and practical application. We treat solution of the differential equation of longitudinal dispersion in porous media (Ogata and Banks 1961) as a similar transformation of the equation for head conduction in solids with strong mathematical boundary conditions. We presuppose for future readers only the background of high mathematics and elementary knowledge of hydrogeology. Three test sites have been used and, namely located in the USA, Germany, and Moldova. Different geological, hydrogeological vulnerability, and pollutants were analyzed. It is definitely clear that new method for groundwater vulnerability assessment is simple, does not need implication of high qualification specialists, and can be applied internationally. In the same time, obtained results correspond to real natural conditions and can be simply verified by the aquifer response in the form of changes of water geochemistry. Proposed methodology is both for local and regional assessments.

Many friends and colleagues have contributed to this book in different ways. Dr. Constantin Moraru extends his deep appreciation to the Council for International Exchange of Scholars, USA for supporting him as a Fulbright Scholar (2013–2014) and Dr., Prof. Robyn Hannigan, Dean, School for the Environment, College of Science and Mathematics, University of Massachusetts Boston for housing him as Fulbright Scholar. Dr. Moraru and Dr. Hannigan have been joined to present the results of this collaborative study in the present publication. The authors also extend their sincere appreciation to Drs. J. Anderson and B. Waldron (Memphis University, USA), Dr. Prof. Hotzl H. (Karlsruhe University, Germany), Dr. Prof. Pozdneakov S. (Moscow State University, Russia), Dr. Prof. Shvets V. (Moscow Geological University, Russia), Dr. Galitskaia I. (Institute for Geoecology, Russia), Dr. Prof. Ryjenko B. (Institute of analytical geochemistry, Russia), Dr. Prof. Haustov A. (International University, Russia), Dr. Prof. StasievGh. (Moldova State University), Dr. Prof. Bulimaga C. (Institute of geography, Moldova), Dr. Prof. Begu A. (Institute of geography, Moldova).

Additionally, a special thanks goes to Olga Moraru, Dr. Moraru's wife, and Dr. Alan Christian, Prof. Hannigan's husband for their patience and support throughout the months we spent writing the book.

Chisinau, Moldova   Constantin Moraru
Boston, USA   Robyn Hannigan

Reference

1. Ogata, A., & Banks, R.B. (1961). *A solution of the differential equation of longitudinal dispersion in porous media.* Geological survey professional paper (441-A, pp. 1–7).

Contents

About the Authors

Dr. hab. Constantin Moraru

<u>Academic Rank</u>
Dr. hab., Head of the Hydrogeology Laboratory, Institute of Geology and Seismology, Academy of Sciences of the Republic of Moldova, Senior Scientific Collaborator.
Degrees, fields, and institutions and dates
BSCE Dnepropetrovsk Mining Academy, Ukraine (1979) Hydrogeology.
Ph.D. Moscow Geological Survey Institute (1987) Hydrogeology and Hydrogeochemistry, Dr. habilitatus of Science, Moscow Institute of Geoecology (2013), Hydrogeology and Geoecology.

Dr. Robyn Hannigan

<u>**Academic Rank**</u>

SMS Professor and Professor at University of Massachusetts Boston; Dean, School for the Environment, College of Science and Mathematics, University of Massachusetts Boston.

<u>**Degrees, fields, and institutions and dates**</u>

BSCE College of New Jersey, Ewing, NJ (1988) Biology.

MSCE MS University of Rochester, Rochester, NY (1994), Geochemistry MA State University of New York at Buffalo, Buffalo, NY (1993) Geology.

Ph.D. University of Rochester, Rochester NY (1997) Geochemistry.

Conversion Factors and Vertical Datum

Concentration of chemical constituents:
mg/l—parts per million (ppm)
µg/l—parts per billion (ppb)
meg/l or epm—milliequivalents per liter or equivalents per million
meg%—percentage of milliequivalents per liter

Multiply	By	To obtain
Cubic meter (m³)	0.0008107	Acre-foot
Kilogram (kg)	2.205	Pound avoirdupois
Kilometer (km)	0.6214	Mile
Megagram (mg)	1.102	Ton (2.000 lb)
Meter	1.094	Yard
Milligram per liter (mg/l)	8.345	Pound per million gallon
Millimeters (mm)	0.03937	Inch
Square kilometers (km²)	0.3861	Square mile

Temperature in degrees Celsius (°C) may be converted to degrees Fahrenheit (°F) as follows: °F = (1.8)°C + 32

Sea level: for the USA refers to the National Geodetic Vertical Datum of 1929
for the Republic of Moldova refers to the former Soviet Union Baltic Datum
for Germany refers to the Deutsches Hauptdreiecksnetz

Specific conductance is given in microsiemens per centimeter of 25 °C (µs/cm at 25 °C).

Concentration of chemical constituents in water are given in either in milligrams per liters (mg/l) or microgram per liter (ug/l).

Well-Numbering System:

(1) for the USA, wells are identified according to the numbering system used by the USGS (US Geological Survey) throughout Tennessee. For example, symbol Sh: U-2 indicates that the well is located in the Shelby County of the "U" quadrangle

and is identified as well 2 in the numeric sequence. Quadrangles are letters from left to right, beginning in the southwest of the county.

(2) for the Republic of Moldova, wells are identified according to the Database of the Institute of Geophysics and Geology of the Academy of Sciences of the Republic of Moldova, 2004.

(3) for Germany, Rastatt area wells are identified according to the database of the Institute of Applied Geology, Karlsruhe University.

Chapter 1
Overview of Groundwater Vulnerability Assessment Methods

Constantin Moraru and Robyn Hannigan

Abstract Contemporary ideas about groundwater vulnerability were discussed. History of the question and the concept of vulnerability are examined. Modern approaches to groundwater vulnerability include usage of overlay, index, statistical, and process-based simulation methods. Each of these methods has been considered in details. The importance and accuracy of the methods are shown in the discussion.

Keywords Groundwater · Groundwater vulnerability
Methods of groundwater vulnerability assessment

1.1 The Concept of Groundwater Vulnerability

The notion **Vulnerability** is originated from the Latin phrase *vulnerabilis, vulnerare* and means easily harmed, hurt, or attacked. From the ancient times of (Egyptian, Roman, and Osman periods), groundwater use has been connected with sever protection and conceptual understanding of aquifer vulnerability to different sources of pollution. For the first time, this notion was attested by Balzac in 1836 (Glossary 2004). Margat (1968) introduced the vulnerability concept in the hydrogeology literature. Albinet and Margat (1971) presented the first map of groundwater vulnerability of the French territory.

Until now, there has been no commonly accepted understanding of the term vulnerability, related to groundwater (Civita 2010; Gurdak 2008; COST actions 620 2003; Witkowski et al. 2007; Zekter et al. 1995; Committee of techniques 1993; Goldberg 1993; Goldberg and Gazda 1984). In general, only groundwater quality is attributed to the practice of vulnerability. Nevertheless, groundwater quantity is also supposed to different vulnerable anthropogenic activities, like overexploitation and sea intrusion. Thus suggesting, in the future, we will need to elaborate one common vulnerability notion, for both groundwater quality and quantity. Furthermore, in most cases, groundwater quality is dependent on quantity and vice versa. In the presented work, we will study groundwater vulnerability related to hydrogeochemical properties or groundwater quality.

© Springer International Publishing AG 2018
C. Moraru and R. Hannigan, *Analysis of Hydrogeochemical Vulnerability*,
Springer Hydrogeology, https://doi.org/10.1007/978-3-319-70960-4_1

Vrba and Zaporozec (1994) proposed the most suitable definition of vulnerability as an intrinsic property of a groundwater system that depends on the sensitivity of the system to human and/or natural impacts. These authors also mention that vulnerability most often is assessed in terms of water quality, and the assessment is based on the uppermost aquifer. Results of the vulnerability assessment are unitless, and they are presented in numerical or logical ratings.

Vrba and Zaporozec (1994) proposed to distinguish intrinsic (natural) and specific types of groundwater vulnerability to pollutants. The authors recognize that such division is arbitrary and not perfect because it is difficult to understand the origin of pollutants. In the context of different types of aquifers and water-bearing rocks, existence Goldscheider (2002) also suggested that distinguishing between intrinsic (natural) and specific (specific land uses and contaminants) is disputable. Civita (2010) also mentions that it is more convenient to divide natural (intrinsic) and specific vulnerability of an aquifer. This author defines that the natural vulnerability depends on three main factors: the ingestion process and the time of travel of water through an unsaturated zone, the groundwater flow dynamics in the saturated zone, and the residual concentration of the contaminant as it reaches the saturated zone. The evaluation of the specific vulnerability of an aquifer should be made case by case, taking into account all the chemical and physical features of every single contaminant.

We also support the idea that division of intrinsic and specific vulnerability is artificial and not arguable, at least from the point of view of modern mass transport pollutant processes. Natural and specific groundwater vulnerability is more professional, and such divisions are appropriate to hydrogeological terminology and understanding. In this publication, we will use the proposed terms.

The existing potential and conditions of aquifer pollution is the main question of the groundwater vulnerability. In this context, the four groups of pollutants are chemical, biological, radioactive, and thermal (Goldberg and Gazda 1984). Following this, classification should be a specific chemical vulnerability, a specific biological vulnerability, etc. Such subdivisions will not complicate hydrogeology terminology, and the practical managerial effect will be significant.

Until now, numerous approaches have been used or proposed for assessing groundwater vulnerability. It is a well-known fact that a single internationally recognized vulnerability assessing method does not exist. In different countries, different methods have been developed and are locally recognized. For example, the DRASTIC method is widely used in the USA (Aller et al. 1987), PI method in Germany (Goldscheider et al. 2000), EPIK method in Switzerland (Practical Guide 1998), and the Goldberg method in Russia and former Soviet Union countries (Goldberg and Gazda 1984). A fundamental characteristic of all approaches for vulnerability assessment is the uncertainty of results. This is a result of characteristics in the method itself or in the data and it uses (Committee on Techniques 1993; Witkowski et al. 2007; Gurdak 2008).

Using different groundwater vulnerability methodologies clearly indicates that this hydrogeological compartment is in the beginning of grounding. At present, an impressionable number of scientific publications exist in this domain. Mostly, all publications are applicable, and only a few of them are dedicated to the theory of question. This phenomenon is explicable by the fact that internationally hydrogeologists are at insufficient levels in high mathematics, physics, and geostatistics. It is paradoxical that hydrogeology itself is one of the numerical sciences between geological disciplines (for instance, hydro geodynamics and hydro geochemistry). In this context, one internationally recognized groundwater vulnerability scale is needed. Analogical situation was, for example, in seismology and stratigraphy. After theoretical background, seismologists have the Richter scale and lithologists have the geologic time scale (GTS) or geochronologic scale.

The primary factors that influence an aquifer's vulnerability to pollutants are summarized in Table 1.1. This is only a general list of factors which are mostly characteristics of the regional distribution of water-bearing horizons. In addition, data from Table 1.1 illustrate that the vulnerability of groundwater consists of complex parameters. For example, the key data in a model used for pesticides transport in soil relating to groundwater pollution consist of 32 parameters, which

Table 1.1 Principal geologic and hydrogeologic features that influence an aquifer vulnerability to contamination (after Johnston 1988 and modified by the authors)

Features determining aquifer vulnerability to contaminants	Low vulnerability	High vulnerability
Hydrogeologic framework		
Unsaturated zone	Thick unsaturated zone, with high levels of clay or organic materials	Thin unsaturated zone, with high levels of coarse sediments such as sand, gravel, limestone, or basalt of high permeability
Confining unit	Thick confining unit of clay or shale above aquifer	No confining unit
Aquifer properties	Silty sandstone or shale limestone of low permeability	Cavernous limestone, sand and gravel, gravel, or basalt of high permeability
Groundwater flow system		
Recharge rate	Negligible recharge rate, as in arid regions	Large recharge rate, as in humid regions
Location within flow system (proximity to recharge or discharge area)	Located in the deep, sluggish part of a regional flow system	Located within a recharge area or within the cone of depression of a pumped well
Management practice		
Wellhead protection	Existing of the well protection zones	No well protection zones
Agriculture	No agricultural practices	Pesticide and fertilizers application
Water abstraction	Water supply from confined units	Water supply from unconfined and karstic aquifers

are combined into 5 groups: pesticide, soil, crop, climatological, and management parameters (Committee on Techniques 1993).

Different classifications of the vulnerability assessment methods have been proposed (Gurdak 2008). We will use the classification proposed by the Committee on Techniques for assessing groundwater vulnerability (USA) (1993). This classification system is more realistic and has been used by many scientists (Civita 2010; Gurdak 2008; Witkowski et al. 2007; COST action 620 2003; Vrba and Zaporozec 1994).

According to the classification (Committee of techniques 1993; Vrba and Zaporozec 1994), assessment methods are placed in three general categories: (a) overlay and index methods, (b) methods employing process-based simulation models, and (c) statistical methods.

1.2 Overlay and Index Methods

Overlay and index procedures are the most popular methods used in the vulnerability assessment practice. Data rely primarily on qualitative and semi-quantitative compilations and interpretations of the maps. In a large spectrum of publications, these methods are described in detail (Civita 1993; Committee on Techniques 1993; Zekter et al. 1995; Corwin and Wagnet 1996; COST action 620 2003; Vrba and Zaporozec 1994; Witkowski et al. 2007; Gurdak 2008; Civita 2010). The DRASTIC (USA), Goldberg method (Russia), and PI (Germany) approaches are representative of this category.

DRASTIC (Aller et al. 1987) is based on the **D**epth to groundwater table, net **R**echarge, **A**quifer media, **S**oil media, **T**opography, **I**mpact of vadose zone, and hydraulic **C**onductivity. Examples of DRASTIC in different geological conditions are presented in Fig. 1.1. The software package calculates the point count system. The higher the vulnerability index is, the greater the groundwater pollution potential is. Mapping of distributed points provides the ability to identify areas that are vulnerable.

The Goldberg method (Goldberg and Gazda 1984) was designed in the former Soviet Union and has been slightly modified by C. Moraru (Jousma et al. 2000). Goldberg's classification is widely used in many countries and is based on variations of four main parameters: lithology of rocks, thickness of low-permeable layers, hydraulic conductivity of rocks, and depth to groundwater tables. Five main classes of vulnerability are recognized from this method: extreme, high, moderate, low, and negligible (Fig. 1.2). Sensitivity analysis points out that the lithology of the unsaturated zone is the main factor governing the vulnerability of groundwater. Goldberg (Goldberg and Gazda 1984) also proposed groundwater vulnerability assessment for confined aquifers. The main idea consists of the relationship between the unconfined water table and the piezometric water level of the confined horizon.

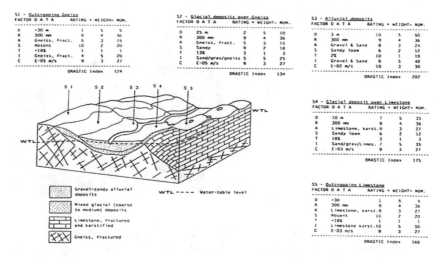

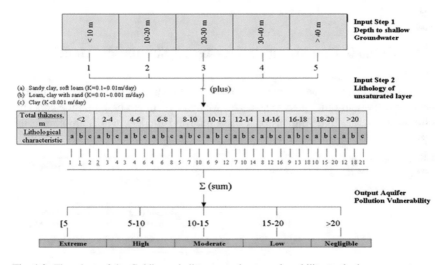

Fig. 1.1 Examples of DRASTIC classification in various hydrogeological conditions (Civita 1990)

Fig. 1.2 Flowchart of the Goldberg shallow groundwater vulnerability method assessment

Goldscheider et al. (2000) proposed the PI method, a Geographic Information Systems (GIS)-based approach to mapping the vulnerability of groundwater contaminants with special consideration of karst horizons. The acronym stands for two factors: protective cover (P) and infiltration conditions (I). The illustration of the PI method is shown in Fig. 1.3 (COST action 620 2003). As shown in Fig. 1.3, the P factor takes into account the effectiveness of the protective cover as a function of the thickness and hydraulic properties of all strata between the ground surface and

Fig. 1.3 Illustration of the PI
method (COST action 620
2003)

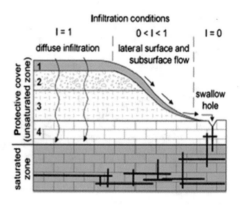

the groundwater surface. The protective cover consists of up to four layers: topsoil, subsoil, non-karst rock, and unsaturated karst rock. The I factor expresses the degree to which the protective cover is bypassed by lateral surface and subsurface flow, especially within the catchment of sinking streams. The simplified scheme of the PI method is presented in Fig. 1.4 (COST action 620 2003).

1.3   Methods Employing Process-based Simulation Models

Process-based simulation methods use the well-known laws of physics, chemistry, geochemistry, and other precise sciences. Many of them are based on the concepts of one-, two-, and three-dimensional mass transport models, which are transferred

Fig. 1.4 Simplified flowchart
for the PI method (COST
action 620 2003)

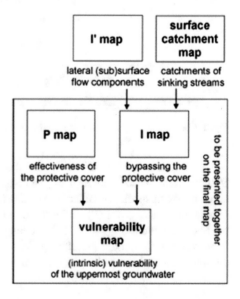

Table 1.2 Major outputs of four types of pesticide simulation models (Committee on Techniques for assessing groundwater vulnerability (USA) 1993)

Output[a]	Type of model			
	LEACHM (Ver.1.0, 1987)	GLEAMS (Ver. 1.8, 54,1989)	PRZM (Release 1, 1985)	CMLS (ver. 4.0, 1987)
Pore-water solute concentration	+	+	+	−
Depth of maximum solute concentration	+	+	+	+
Maximum depth of solute penetration	+	+	+	−
Soil-water flux	+	+	−	−
Soil-water content	+	+	+	−
Phase partitioning of solute mass	+	+	+	−
Temperature	*	*	−	−
Water uptake	+	+	+	−
Pesticide uptake	*	+	*	−
Volatile losses	+	−	−	−
Runoff	−	+	+	−

[a]A plus sign indicates that the output parameter values are provided, and a minus sign indicates that the parameter values are not provided. An asterisk indicates that the parameter can be considered, but usually is not because of insufficient data

in geology from the section—transport of heat and electricity, and physics of solids. Process-based models can be used both in regional and in site studies. Usually, such models need complex data, which are not easily available. Table 1.2 indicates the major outputs of four models of pesticide simulation. Data from this table have shown the main capabilities of models and final results. Such models are used in two different ways (Focazio et al. 2002):

1. To determine the intrinsic susceptibility of an aquifer
2. To assess the vulnerability of the groundwater supply to a targeted contaminant or contaminants.

Also process-based simulation methods are described by Holtschlag and Luukkonen (1997), Levy et al. (1998), Connel and van den Daele (2003), Gurdak (2008). Holtschlag and Luukkonen (1997) used the pesticide-leaching model, which is characteristic and representative of process-based simulation models. These authors describe pesticide leaching by a steady-state unsaturated zone transport model. The model calculates fractions of pesticide remaining in the unsaturated zone and the time of travel in the same zone as function of depth. The model itself contains the following assumptions:

1. Water content in the unsaturated zone may be described as a static function of elevation above the water table;
2. The water flux is steady, downward, and diminished linearly with depth in the root zone;
3. All water in the unsaturated zone transports pesticides;
4. Water initially present in the profile is completely displaced downward by water entering from above;
5. Pesticides are in aqueous phase or are absorbed;
6. Absorption is described by a linear, reversible equilibrium relation;
7. Pesticide decay is an irreversible first-order reaction;
8. Pesticides occur at concentrations small enough that neither the capabilities for absorption nor decay of the pesticides by materials in the unsaturated zone are exhausted;
9. Pesticide loss in the root zone is proportional to the root uptake of water and subsurface runoff;
10. Pesticides are applied at the land surface at a constant rate, pesticide flux is a steady state everywhere, and dispersion of pesticide is negligible.

After all assumptions, the criterion for vulnerability is the fraction of atrazine remaining at the water table. The model was applied to data of 5,444 wells in Kent County, Michigan, USA (Holtschlag and Luukkonen 1997). It is necessary to outline that proposed model has to sever assumptions. In many cases, it is difficult to maintain all theoretical conditions. This leads to probability decreasing of modeled results.

1.4 Statistical Methods

Statistical methods are not often used for vulnerability assessment. Nevertheless, since groundwater vulnerability is a probabilistic notion, statistical methods should have more application in vulnerability assessment (Committee on Techniques 1993). This publication states that two practical applications of statistical techniques in vulnerability assessments are possible:

1. regionalization—differentiating of groundwater regions and
2. assessment of vulnerability with probability models—such as principal component, cluster, time series, and regression analyses.

Vulnerability assessment statistical methods have been applied by Teso et al. (1996), Nolan (2001), Voss (2003), and Gurdak (2008). Generally, these methods use different geostatistical parameters and procedures, such as functional relationships, cluster and component analyses, multivariate regressions, and simple parametrical statistics. For example, the Voss (2003) reliability of the atrazine map was assessed by using statistical procedures.

1.5   Discussion

It is evident that Groundwater Vulnerability Assessment (GWVA) became a new discipline of hydrogeology, which combines knowledge from geology, general hydrogeology, hydrogeochemistry, geochemistry, geostatistics, and physics of mass transport phenomena. As a new discipline, GWVA is in the period of methodology development and data collection. GWVA existing methods do not have international circulation and application. In addition, internationally recognized method(s) by all professional community do not exist. Practically, all analysed methods assess vulnerability of the first aquifer from the land or unconfined surfaces. Nevertheless, hydrogeological system includes unconfined and confined aquifers. In many countries, confined aquifers serve for freshwater supplies of economy and society.

Overlay and index methods were the first applied in hydrogeology and are the most simple. They are based on the classical principle of geology—map overlaying or superposing a series of maps. The main concern in this action is working in the same topographical scale. As a rule, the final product is a single complex map, which in some cases, is difficult to read. Index methodology assigns a numerical value to each proposed or selected parameter.

For overlay and index methods, the magnitude of the rating is not always perfectly argued. In many classifications, this rating is logical or comparative. In only a few methods (e.g., Goldberg method), rating system is connected with hydrogeological properties of rocks. Only Goldberg method is applied both for shallow (unconfined) and for deep (confined) aquifers. The rating results can be converted in numerical migration time values of the pollutant (Goldberg 1987; Goldberg and Gazda 1984). Five vulnerability classes are equivalent to the pollutant migration time factor (t) in the following relations: 1 class (extreme)—$t < 10$ days; 2 class (high)—10 days $< t < 50$ days; 3 class (moderate)—50 days $< t < 100$ days; 4 class (low)—100 days $< t < 200$ days; 5 class (negligible)—$t > 200$ days.

Methodologically, GWVA for confined aquifers is still scarce in application. Practically, only Goldberg method includes the vulnerability evaluation for such type of aquifers. Three parameters characterize the hydraulic relationship between confined and unconfined aquifers and, namely m_o is the thickness of impermeable stratum between aquifers (m), H_1 is the water level in unconfined aquifer (m), and H_2 is the water level (m) in confined one. Three groundwater vulnerability classes may be mentioned using these parameters:

- *1st class* (protected)—$m_o > 10$ m and $H_2 > H_1$;
- *2nd class* (conventionally protected)—

  case (a) $5 < m_o < 10$ m and $H_2 > H_1$
  case (b) $m_o > 10$ m and $H_2 < H_1$;

- *3d class* (unprotected)—$m_o < 5$ m and $H_2 > H_1$ or $H_2 < H_1$.

Time of pollution migration is proposed for these classes. The values of time change from 1 year to more than 20 years depending on values of m_o and hydraulic conductivity of the impermeable stratum.

Process-based simulation methods contain numerical and theoretical argumentation. Nevertheless, they have many limitations (mathematical, physical, and logistical), and results are as good as the smallest investigated territory. Data collection and assessment are also problematic for these models, which can lead to a major source of error. Initial conditions for such models are very strong and not realistic. The model of Holtschlag and Luukkonen (1997) contains ten assumptions. One of them stated "Pesticides are applied at the land surface at a constant rate, pesticide flux is a steady state everywhere, and dispersion of pesticide is negligible." Such conditions are difficult to maintain in a laboratory environment and are unpractical in real agricultural situations. The convective–dispersive solute transport approach, employed in the model LEACHM and others of this type, predicts the asymptotic behavior and is least likely to be valid when used for shallow depths (Committee on Techniques 1993). Calibrating the model parameters using experimental data from the zone of interest can circumvent this problem, but extrapolation to much greater depths in the vadose zone can lead to significant errors.

Quality of statistical methods depends on the accuracy of initial data. Results of such methods should be interpreted in their real connection with geological, hydrogeological, and environmental conditions. For the territory of Moldova, cluster analysis permits distinguishing vulnerable chemical types of groundwater (Moraru et al. 1990). For the High Plains regional aquifer (USA), a methodology was tested. It was proposed to quantify prediction uncertainty with multivariate logistic regression using Geographical Informational System. Latin hypercube sampling is applied to illustrate the propagation of impute error and estimate uncertainty associated with the output of the logistic-based groundwater vulnerability model (Gurdak 2008).

Very often different methods have shown various results for the same territory. This question was analyzed for some European countries (COST action 620 2003). The Sierra de Libar territory (Spain) is a representative example, and PI (Goldscheider et al. 2000) and COP (Vias et al. 2002) methods were used (Figs. 1.5 and 1.6).

Statistical uncertainty is a characteristic for all groundwater vulnerability models. It should be noted that the uncertainty is well studied and quantifying in physics and mechanics (the internationally recognized principle of uncertainty proposed by Werner Heisenberg). Uncertainties are different by notion and background in mathematical sciences and geology. In geology, the term uncertainty is used more

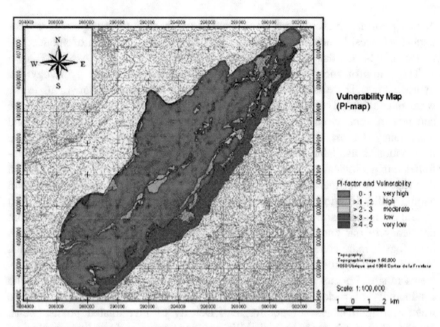

Fig. 1.5 Vulnerability map by PI method of the Sierra de Libar (COST action 620 2003)

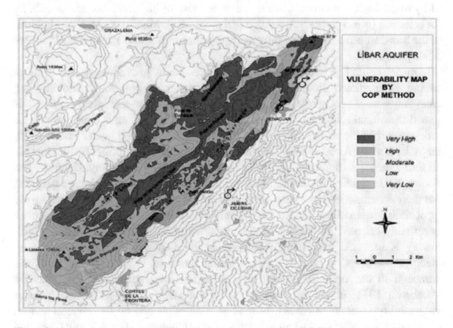

Fig. 1.6 Vulnerability map by COP method in Sierra de Libar (COST action 620 2003)

or less informal. Nevertheless, consideration of uncertainty is important in geological sciences. In details, this question is analyzed in a spectrum of publications (Hancen 2014; Gurdak 2008; Bardossy and Fodor 2001; Kundzewicz 1995).

The notion of uncertainty is still in scientific discussion in the hydrology and hydrogeology. There is a plethora of single words which are synonymous to the word uncertainty. The meaning of the term uncertainty partly overlaps with the contents of such words as doubt, dubiety, skepticism, suspicion, mistrust, and inconstancy. Uncertainty is obviously opposed to certainty, where the complete information is available (Kundzewicz 1995). Uncertainty and error may be confounded in geological studies. Uncertainty, in the broadest sense, is the recognition that results of our measurements and observations may deviate more or less from natural reality (Bardossy and Fodor 2001). Uncertainty in hydrology results from natural complexity and variability of hydrological systems and processes and from deficiency in our knowledge (Kundzewicz 1995). Hydrogeology compared with hydrology is more complex related to data collection and parameter measurements. Practically, all hydrogeological data are obtained indirectly without real measurements in situ. For instance, groundwater movement is calculated using hydraulic conductivity of rocks; rock water transitivity is calculated using probabilistic water-bearing thickness of aquifer, etc. Density of investigated points (wells, springs, etc.) is the other problem. We can only navigate from hydrogeological point-to-point with assumption that geological media is homogeneous. The number of hydrogeological points is always scarce because of high cost and inaccessibility for selection (urban territories, economical areas, etc.). Moreover, geological points, in the broadest sense, are three-dimensional (coordinate X and Y, and Z is altitude or other parameter).

Among natural geological natural resources, only groundwater can refill or restore accumulated liquid reserves. Groundwater is constantly moving and changing its geochemical properties. Each hydrogeological point is four-dimensional (classical X, Y, and Z, and t, time). The same is characteristic for groundwater vulnerability assessment (for simplicity, a groundwater vulnerability map). In this case, conditions of uncertainty (unexpected and unforeseeable) include all four dimensions and such system became more complex for study. Bardossy and Fodor (2001) mention that four parametric calculations are more and more frequent, but their uncertainties have been treated so far only by deterministic and stochastic (probabilistic) methods. Also these authors put into discussion the main uncertainty-oriented methods with a short description of each of them. The following is included and, namely interval analysis, possibility theory, fuzzy set theory, the probability bounds method, neural networks, and the method of hybrid numbers. In our opinion, geostatistics or regionalized spatial variables method (Matheron 1971) is a good instrument for uncertainty analysis of groundwater vulnerability assessment results. Much has been published on the method and on practical experience (Isaaks and Srivastava 1989; Clark and Harper 2000; Chun and Griffith 2013). At present geostatistics became a powerful method, and it is widely used in geological science and practice. The conditions for properly posing a method are the following.

(a) For uncertainty analysis of the groundwater vulnerability, the main attention and efforts are proposed to be focussed on: selection of a method for future groundwater vulnerability assessment that needs to be correlated with the importance of study territory, geological and hydrogeological conditions and purpose of final results. The main tendency for current groundwater vulnerability models is to cover big territories (country, region, etc.). It is well known that inside of such territories exist a great many areas which do not present interest in vulnerability assessments because groundwater is not used for water supply (i.e., forest area and big factories). Other important processes are the process of self-purification or natural remediation of the aquifer need to be taken into account. Moraru (2009) investigated strong nitrogen groundwater pollution under the influence of livestocks. It was modeled and confirmed by sampling that nitrogen pollution (as NO_3 and NO_2) migrated about 1800 m from the sources. In this interval of migration, self-groundwater purification processes reduced nitrogen pollution to natural concentration. Moraru et al. (2005) demonstrated the same idea for the shallow groundwater of the Republic of Moldova.

Methodologically, this principle of selection is used in many geological applications. Representative example is seismology. For instance, we have an earthquake with magnitude 7.5, and two territories are affected. One is a city and other is a desert area. Seismological zonation of the desert territory does not make sense because it is without buildings and social infrastructure. The same situation is with groundwater. Only special places are used for water extraction. It is important to outline that geochemical migration of pollutants is limited in distances and it is regulated by the aquifer natural proprieties.

(b) Data collection should take into account errors occurring during sampling, observations, and measurements. All collected hydrogeological data contain errors which can be divided formally into two categories: constant errors and occasional or random errors. Constant errors usually are connected with metrological characteristics of instruments, professionalism of operators, climatic conditions, etc. In other words, such errors are detectable and have simple statistical properties. Occasional or random errors do not have predictable statistics, and in many cases, we can investigate them only in statistical intervals. Very frequently random errors remain uncorrectable (for instance, data from a well plumbing test and geochemical composition of deep groundwater).

(c) Mapping models and its accuracy need special attention. This question is well studied, but specialists do not spare sufficient attention. Depending on the objective of study, vulnerability maps are constructed in zones (general information and limited points) and in isolines (detailed mapping and accurate). A large spectrum of algorithms exists for data interpolation (Surfer, MapInfo, ArcGIS, etc.). Mostly, all of them are based on known spacial interpolators. Moraru and Timoshencova (2013) studied hydrogeological map accuracy. Maps were modeled in isolines by means of different interpolators. The occurrence of groundwater levels was mapped on the representative area in limits of the Republic of Moldova. Using the software Surfer 11, eleven models of groundwater levels were generated by the following interpolators: Kriging,

Radial Basis Function, Inverse Distance to a Power, Modified Shepard's, Minimum Curvature, Polynomial Regression, Triangulation with Linear Interpolation, Nearest Neighbor, Natural Neighbor, Moving Average, and Local Polynomial. The comparability between maps was estimated applying methodology of the statistical residuals as the subtraction between natural model/artificial model and the absolute average value of the statistical residuals. Modified Shepard's method is defined as the most accurate interpolator. Nevertheless, the correlation analysis shows that this interpolation method is very close with Kriging, Local Polynomial, Minimum Curvature, Natural Neighbor, and Triangulation with Linear Interpolation ($r > 0.70$).

(d) Variograms are suitable to study spatial variability, spatial correlation of variables and to determine their ranges of influence (Bardossy and Fodor 2001). From the first groundwater vulnerability map (Margat 1968) until the current avalanches of publications, one principal question is at the top of the professional discussions: That is the permitted accuracy of distance between interpolated points on the map? In other words, this means how homogenous is groundwater media between observations points (wells, springs, etc.). In mining geology, this question is important because of the real practical implementation (Rock 1988). In our case, special variability of data and aquifer homogeneity is directly connected with validation of the constructed model. In such conditions, variogram (s) analysis is a very powerful tool.

(e) Other mathematical and statistical methods can by used. Cluster analysis permits the understanding of relationships between groundwater vulnerability groups and to check their geostatistical similarity. Component analysis is similar to cluster method. Monte Carlo analysis gives good results and is understanding by geologists. These methods consist of computer-based calculations and models (e.g., computer programs SPSS 14 and Statistica).

References

Albinet, M., & Margat, J. (1971). Cartographie de la vulnerabilite a ala pollution des nappes d'eau souterraine. Ground water pollution symposium. In: *Proceedings of the Moscow Symposium, August 1971. Actes du collogue du Moscow. Aout 1971): IASH – AISH Publ. No.*103.

Aller, L., Bennett, T., Lehr, J. H., & Petty, R. J. (1987). *DRASTIC: A standardized system for evaluating ground water pollution potential using hydrogeological settings.* Environmental Protection Agency, Oklahoma: U.S.

Bardossy, G., & Fodor, J. (2001). Traditional and new ways to handle uncertainty in geology. *Natural Resources Research, 10*(3), 179–187.

Chun, Y., & Griffith, D. A. (2013). Spatial statistics and geostatistics: theory and applications for geographic information science and technology (SAGE advances in geographic information science and technology series) (200p). California: SAGE Publications Ltd.

Civita, M. (1990). La valutazione della vulnerabilita degli acquiferi all'inquinamento. Proc. 1st Conv. Naz. "Protezione e Gestione delle Acque Sotterranee: Metodologie, Technologie e Obiettivi". *Marano sul Panaro, 3,* 39–86.

Civita, M. (1993). Ground water vulnerability maps: a review. In *Proceedings IX Symposium on Pecticide Chemistry, Degradation and Mobility of Xenobiotics, Piacenza, Italy, Lucca (Biagini)*1993 (pp. 587–631).

Civita, M. (2010). The combined approach when assessing and mapping groundwater vulnerability to contamination. *Journal Water Resources and Proyection, 2010*(2), 14–28.

Clark, I., & Harper, W. V. (2000). *Practical geostatistics* (442p). Ecosse North Amer Llc.

Committee on Techniques for assessing ground vulnerability (USA). (1993). Ground water vulnerability assessment: contamination potential under conditions of uncertainty (204p). Washington: National Academic Press.

Connel, L. D., & van den Daele, G. (2003). A quantitative approach to aquifer vulnerability mapping. *J. of Hydrology, 276,* 71–78.

COST action 620. (2003). In F. Zwahlen (Ed.) *Vulnerability and risk mapping for the protection of the carbonate (karst) aquifer* (297p). Final report. European Commission, Directorate—General for Research.

Focazio, Michel, L., Reilly, T. E., Rupper, M. G., & Helsel, D. R. (2002). Assessing ground water vulnerability to contamination: Providing scientifically defensible information for decision makers. U.S. Geological Circular 1224, 33 p.

Glossary. (2004). http://webword.unesco.org/water/ihp/db/glossary/ http://webworld.unesco.org/water/ihp/db/glossary/glu/aglu.htm.

Goldscheider, N. (2002). *Hydrogeology and vulnerability of karst systems—examples from the Northen Alps and Swabian Alb.* Ph.D. thesis, Institute of applied geology, University of Karlsruhe, Germany. 236 p.

Goldscheider, N., Klute, M., Sturm, S., & Hotzl, H. (2000). The PI method—a GIS-based approach to mapping ground water vulnerability with special consideration of karst aquifer. *Zeitschnft Angewandte Geologie, 46*(3), 153–166.

Goldberg, V. M. (1993). Natural protection of groundwater against contamination. In Y. Eckstein & A. Zaporozec (Eds.) *Proceedings, Second USA/Cis Joint Conference on Environmental Hydrology and Hydrogeology, Washington, DC. Water management and protection* (pp. 141–145). Alexandria, Virginia: American Institute of Hydrology.

Goldberg, V. M., & Gazda, S. (1984). *Gidrogeologicheskie osnovy okhrany podzemnykh vod ot zagryazneniya* (Hydrogeological principles of groundwater protection against pollution) (239p). Moscow: Nedra.

Goldberg, V. M. (1987). Vzaimosveazi zagreaznenia podzemnyh vod i prirodnoi sredy (248p). Moscow: Gidrometeoizdat.

Gurdak, J. J. (2008). Ground-water vulnerability: Nonpoint-source contamination, climate variability, and the High Plains aquifer (223p). Saarbrucken, Germany: VDM Verlag Publishing. ISBN: 978-3-639-09427-5.

Hancen, D. T. (2014). http://proceedings.esri.com/library/userconf/proc98/proceed/to200/pap183/p183.htm.

Holtschlag, D. J., & Luukkonen C. L. (1997). *Vulnerability of ground water to atrazine leaching in Kent County*, Michigan. U.S. Geological Water—Resources Investigations Report 96-4198: 49 p.

Isaaks, E. H., & Srivastava, R. M. (1989). *Applied geostatistics (561p).* Oxford: Oxford University Press.

Johnston, R. H. (1988). Factor affecting ground water quality. National water summary 1986: Hydrologic events and ground water quality. Water-Supply paper. Reston, Virginia. U.S. Geological Survey 2325: 32 p.

Jousma, G., Kloosterman, F., Moraru, C., et al. (2000). Groundwater and land use. Report of the TACIS Prut water management project: 180 p.

Kundzewicz, Z. W. (1995). Hydrological uncertainty in perspective. In Z. W. Kundzewicz (Ed.), *New uncertainty concepts in hydrology and water resources* (pp. 3–10). Cambridge: Cambridge University Press.

Levy, J. l., et al. (1998). Assessing aquifer susceptibility to and severity of atrazine contamination at field site in south-central Wisconsin, USA. *Hydrogeology Journal, 6,* 483–499.

Margat, J. (1968). Vulnerabilite des nappes d'eau souterraines a la pollution. Bases de la cartographie. BRGM# 68. SLG 198 HYD. Orleans.

Matheron, G. (1971). *The theory of regionalized variables and its applications* (211p). Ecole nationale supe rieuredes mines, Paris.

Moraru, C. E. (2009). Gidrogeohimia podzemnyh vod zony activnogo vodoobmena krainego Iugo-Zapada Vostocno—Evropeiskoi platformy. Chisinau: Elena V.I.: 210p.

Moraru, C., Budesteanu, S., & Jousma, G. (2005). Typical shallow groundwater geochemistry in the Republic of Moldova (pilot study). *Buletinul Institutului de Geofizica si Geologie al Academiei de Stiinte a Moldovei, nr., 1,* 36–48.

Moraru, C. E., Burdaev, V. P., & Negrutsa, P. N. (1990). Classification and evaluation of hydrogeochemical facies using the cluster analysis. In *Deposited with VINITI,* 1990, No 6497-V90, Moscov: 15p.

Moraru, C. E., & Timoshencova, A. N. (2013). Evaluation of spatial interpolation methods for groundwater (case study, the Republic of Moldova). *Buletinul Institutului de Geofizica si Geologie al Academiei de Stiinte a Moldovei, nr., 1,* 24–42.

Nolan, B. T. (2001). Relating nitrogen sources and aquifer susceptibility to nitrate in shallow ground water of the United States. *Ground Water, 39*(2), 290–299.

Practical guide. Ground water vulnerability mapping in karstic regions (EPIK). (1998). Swiss Agency for the Environment, Forests and Landscape (SAEFL): 56p.

Rock, N. M. S. (1988). Lecture Notes in Earth Sciences. In S. Bhattacharji, G. M. Friedman, H. J Neugebauer, & A. Scielacher (Eds.) *Numerical geology* (427p). Berlin: Springer.

Teso, R. R., et al. (1996). Use of logistics regressions and GIS modelling to predict groundwater vulnerability to pesticides. *Journal of Environmental Quality, 25,* 425–432.

Vias, J. M., et al. (2002). Preliminary proposal of a method for vulnerability mapping in carbonate aquifers. In F. Caraso, J. J. Duran, & B. Andreo (Eds.) *Carst and environment* (pp. 75–83).

Voss, F. D. (2003). Development and testing of methods for assessing and mapping agricultural areas susceptible to atrazine leaching in the State of Washington. U.S. Geological Survey Water—Resources Investigation Report 03-4173: 13 p.

Vrba, V., & Zaporozec, A. (1994*). Guidebook on mapping ground water vulnerability. International association of hydrogeologists* (vol. 16, 90 p).

Witkowski, A. J., Kowalczyk, A., & Vrba, J. (2007). Groundwater vulnerability assessment and mapping. In: selected papers from the groundwater vulnerability assessment and mapping conference. Ustron, Poland, 2004 (263p). London, UK: Taylor and Francis group.

Zekter, I. S., Belousova, A. P., & Yu, Dudov V. (1995). Regional assessment and mapping of groundwater vulnerability to contamination. *Environmental Geology, 25,* 225–231.

Chapter 2
Geochemical Method of the Groundwater Vulnerability Assessment

Constantin Moraru and Robyn Hannigan

Abstract The new geochemical method of groundwater vulnerability assessment is proposed. In order to background this method, longitudinal geochemical profile is referred as geochemical signals and theoretical background of pollutant migration in unsaturated zone is analyzed. Final point of geochemical migration in the unsaturated zone has been argued. Three independent methods of the final point of migration are discussed, and namely laboratory, statistical, and experimental. On the basis of this, geochemical aquifer vulnerability estimation leakage potential methodology (GAVEL) is justified.

Keywords Geochemical signals · Pollutant migration · Point of migration Unsaturated zone

2.1 Geochemical Signals in the Unsaturated Zone

The term "signal" is derived from the Latin "signium," meaning sign. Any time-variable physical phenomenon that can convey information is called a signal (Bhadoria 2013). The concept of signals is widely used in physics (electromagnetic signals, etc.), chemistry (signals of chemical reaction, etc.), biology (coded messages sent from one organism to another), and other sciences. In geology, signals are used in geophysics, paleontology, structural geology, and hydrogeology, among other disciplines. In geochemistry, the term "signal" is not commonly used. Nevertheless, the study of geochemical signals is important in applied geochemistry because of their relevance to mineral deposits and special techniques for data interpretation (Krainov 1973; Perelman 1982, 1989; Geohimiceskie metody poiskov rundyh mestorojdenii 1982; Faure 1998).

For our study, we will define signals as changing elementary chemical concentration in geological media relative to existing natural (background) concentration. In other words, disturbance of natural geochemical equilibrium leads to

© Springer International Publishing AG 2018
C. Moraru and R. Hannigan, *Analysis of Hydrogeochemical Vulnerability*,
Springer Hydrogeology, https://doi.org/10.1007/978-3-319-70960-4_2

appearance of geochemical signals. The unsaturated (aeration or vadose) zone is suitable as a major repository of geochemical signals. Moraru (2009), Gurdak (2008), Edmunts and Tyler (2002), Cook et al. (1992), and Perelman (1989) describe and investigate different peculiarities of geochemical signals in the unsaturated zone. In this zone, geochemical signals are the indicators of climate variability, paleorecharge and paleoclimate, surface pollution migration, and contamination (or pollution) approaching the unconfined aquifer.

In the unsaturated zone, geochemical signals are theoretically created by chemical elements or chemical compounds. Fluctuation of concentration (C) in point(s) during time (t) is $C = f(t)$ or $C = f(X, Y, t)$ along depth (h) with coordinates (X, Y). Physically, these fluctuations are among the family of signals. An example is shown in Fig. 2.1.

It is important to describe geochemical signals bearing in mind the essential fundamentals of signal classification (Priemer 1991; Kay 1993). Theoretically, geochemical signals should be continuous in time because of convection or dispersion in porous media. However, it is impractical to define geochemical information at every instant of time. For this reason, we will consider geochemical signals to be discrete—the signal can be defined as data at a discrete instant of time. In this approximation, geochemical signals are digital. Usually, all signals can be periodic or aperiodic relative to time. If a signal is repeated after uniform amount of time, it is defined as periodic. Otherwise, the signal is defined as aperiodic. In most cases, geochemical signals are aperiodic (see Fig. 2.1).

The unsaturated zone is exposed to surface pollutants inconstantly and aperiodically (despite point and nonpoint sources of contaminates). This characteristic accounts for the variability in a geochemical signal. Random signals can be treated with a wide spectrum of mathematical functions. Depending on the purpose of assessment, these functions may be applied to the entire length of a geochemical signal or a segment of it. If a line (ecological limits line, maximum admissible concentration, etc.) divides a geochemical signal, it becomes even and odd sub-signals. Also geochemical signals can be amplitude and time scaling.

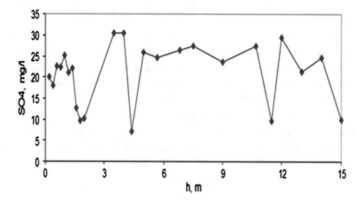

Fig. 2.1 Geochemical signal of sulfate (SO$_4$) in the unsaturated zone as function of depth (h)

In the unsaturated zone, sources of surface pollution are irregular and dependent on many factors. In porous media, transportation of a pollutant occurs with water flux. Water originates from atmospheric precipitation (rain or melted snow), inundations, and anthropogenic accidents. Geochemical signals have impulse properties in continuous time periods. This principle is important, because all applied solutions of the differential equation of longitudinal dispersion in porous media are based upon it.

2.2   Theoretical Background of Pollutant Migration in the Unsaturated Zone

Movement or migration of a pollutant in the unsaturated zone can be described by the simplest longitudinal dispersion model, which was initially proposed by Radushkevich (1960) and Taylor (1954). Ogata and Banks (1961) continued this investigation and proposed a solution to the differential equation of longitudinal dispersion in porous media. This proposal is used in many practical geochemical applications concerning the unsaturated zone. Shestakov (1973), Pashkovskii (1973), and Rosali (1980) developed different interpretations of and substantial additions to the differential equation of longitudinal dispersion in porous media.

All assumptions involving the migration of pollutants in the unsaturated zone are based upon differential equations of conduction (transferring) of heat in solids. From this theory, basic differential equations were adapted to geochemical conditions of the unsaturated zone. In order to understand the essentials of the parallelism between conduction of heat in solids and pollution migration in the unsaturated zone, we will analyze one simple example. Carlslaw and Jaegar (1959) describe the mathematics behind this representative example.

Consider an isotropic solid, like a rod with length $l > 0$. Initial temperature of the solid (at the tip of the rod) is $T = 0$. At time $t > 0$, we change the temperature at the tip of the rod to $T_1 > T$ for a limited period of time $t = t_1$. The changing temperature (T_i) of the rod can be described as a function incorporating coordinates (X, Y, and Z) and time (t) accordingly:

$$T_i = f\left(X, Y, Z, t_1\right)$$

Question: What is the distribution of temperature in the entire rod for the time $t = t_1$? According to the fundamental principles of heat conduction (Carlslaw and Jaegar 1959), the answer to the question is best represented graphically (Fig. 2.2).

Analysis of Fig. 2.2 suggests the following ideas. If the temperature input to the rod (solid body) occurs over a limited period of time ($t = t_1$), only a portion of the rod is affected by changes to the initial temperature. In such conditions, the temperature distribution is an exponential function for the length l_1 only. After the interval ($0.0 - l_1$), initial temperature is unchanged. In other words, conduction of

Fig. 2.2 Temperature distribution in the solid rod (l_1 is a part of l for which $T_1 > T$; the rest of the abbreviations are found in the text)

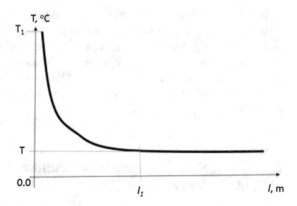

temperature is finite in an isotropic solid for an impulse period. Thermal conductivity is a regulator of this process.

The described thermal process is precisely analogous to Ohm's law for the steady flow of electric current: The flux f corresponds to the electric current, and the fall in temperature $T - T_1$ corresponds to the fall in potential. This is also because of the existing electrical resistance (Carlslaw and Jaegar 1959).

This idea is applied to the simplest longitudinal dispersion model for pollutants in porous media, including the unsaturated zone (Radushkevich 1960; Taylor 1954; Ogata and Banks 1961; Shestakov 1973; Pashkovskii 1973; Rosali 1980). In present-day models, many scientists fail to take this critical idea into full consideration. Consequently, modeled results do not always correspond to natural conditions.

We will analyze in more detail the publication (Ogata and Banks 1961) because of its popularity and practical application of the proposed solution of the differential equation of longitudinal dispersion in porous media. Before solving the differential equation of dispersion, the authors (Ogata and Banks 1961) proposed the following strong conditions:

1. The porous medium is homogenous and isotropic.
2. No mass transfer occurs between solid and liquid phases.
3. Due to microscopic velocity variations in the flow tube, solute transport across any fixed plane may be quantitatively expressed as the product of dispersion and the concentration gradient.
4. The flow in the medium is unidirectional, and the average velocity is taken to be constant throughout the length of the flow field.

The problem was to characterize the concentration of the flow (C) as a function of migration distance (x) and time (t) using Eq. 2.1 (Ogata and Banks 1961):

$$C(x,t) = \Gamma(x,t) = \exp\left(ux/2D - u^2t/4D\right) \qquad (2.1)$$

where

Γ gradient of concentration,
u average velocity of fluid in porous medium,
D dispersion coefficient.

Equation 2.1 is as if the same applied for conduction of heat in solids (Carlslaw and Jaegar 1959). For this reason Ogata and Banks (1961) utilized Duhamel's theorem (Carlslaw and Jaegar 1959), and for specified boundary conditions finally obtain Eq. 2.2:

$$C/C_0 = 0.5 \, [\text{erfc} \{(1- \varepsilon)/2\sqrt{\varepsilon\mu}) + \exp (1/\mu) \, \text{erfc} \, (1+ \varepsilon)/2\sqrt{\varepsilon\mu})] \qquad (2.2)$$

where

$\varepsilon = ut/x$ and $\mu = D/ux$ and C_0—concentration in the source,

erfc   complementary error function (or Gaussian invert function),
exp   exponential function.

The properties of Eq. 2.2 have been analyzed by Ogata and Banks (1961). The main results are as follows:

1. If a point at great distance away from the source is considered, then it is possible to approximate the boundary conditions by $C(-\infty, t) = C_0$, which leads to a symmetric solution.
2. Graphical representation of the function $\mu = f\,(C/C_0)$ shows that as μ becomes small the concentration distribution becomes nearly symmetric about the value $\xi = 1$.
3. This indicates that for large values of D or small values of distance x, the contribution of the second term in Eq. 2.2 becomes significant as ξ approaches unity.
4. Experimental testing presents evidence that the distribution is symmetrical for values of x chosen some distance from the source.

This simple model is often still applied to the unsaturated zone. Rosali (1980) proposed a practical application of the differential model. The initial equation is presented in a Fickian-type form:

$$n(\partial\hat{C}/\partial t) = D(\partial^2\hat{C}/\partial x^2) - v(\partial\hat{C}/\partial x) \qquad (2.3)$$

where

n	effective porosity, fraction,
$n = n_a + 1/\beta; n_a$	the active porosity and β is the absorption coefficient,
$\hat{C} = (C - C_o)/(C^\circ - C_o)$ and C°	pollution concentration at the entrance into the medium, mg/l,
C_o	pollution concentration (background) in the unsaturated zone, mg/l,
C	estimated concentration of a pollutant, mg/l,

D	coefficient of longitudinal dispersion, m^2/day,
v	linear velocity of the liquid flow, m/day,
t	time, day,
x	distance from the point of release, m.

The solution of Eq. 2.3 for the following conditions

$$C(x, o) = C^o$$
$$C(o, t) = C^o$$
$$C(\infty, t) = C_o$$
$$vx/D > 0.1$$

is as follows (Rosali 1980):

$$\widehat{C} = (C - C_o)/(C^o - C_o) = 0.5\left\{\mathrm{erf}(\xi) + \exp(\mu^2)\mathrm{erf}\left(\xi^2 + \mu^2\right)^{1/2}\right\} \qquad (2.4)$$

where

$$\xi = \left\{(nx - vt)/(2Dnt)^{1/2}\right\}$$

$\mu = vx/DS$; S—area of infiltration, m^2.

As Radushkevich (1960) and Ogata and Banks (1961) indicated, the second term of Eq. 2.4 becomes very low at $t \to \infty$ and it can be expressed in the reduced form:

$$\widehat{C} = (C - C_o)/(C^o - C_o) = 0.5\{\mathrm{erfc}(\xi) \qquad (2.5)$$

where

erfc   the Gaussian invert function

If we determine the value C (estimated concentration of a pollutant) from Eq. 2.5, the result is as follows:

$$C = (C^o - C_o)\,0.5\,\{\mathrm{erfc}(\xi) + C_o \qquad (2.6)$$

Since C is dependent on time t as $C = f(t)$ and for $t = t_1 =$ constant, the solution of Eq. 2.6 shows that at depth x exists condition $C = C_o$. Graphically, results are similar to those in Fig. 2.2. This means that in the relation to the unsaturated zone, migration of pollutants is not continuous. Depending on the concentration gradient and its time activity t, depth of pollutant migration is quasi-final. In this context, we propose the term quasi-point of pollutant migration PM in the unsaturated zone. Equation 2.7 determines the PM (l_k) as

$$l_k = x(C, t) \tag{2.7}$$

2.3   Numerical Determination of the PM

Analysis of Eq. 2.3 in the form of Eq. 2.5 shows that if $t = m$ (where m is a fixed value of time), pollutant migration in the unsaturated zone is finite. In other words, pollutants penetrate to a fixed depth only. The value of n as the effective porosity takes into account the absorption of a pollutant by the matrix of the unsaturated zone. In this case, Eq. 2.5 becomes:

$$\left(C_{(x,t)} - C_o\right)/\left(C^o - C_o\right) = 0.5\,\mathrm{erfc}\left[\left(nx - vt/\left(2(2Dnt)^{1/2}\right)\right)\right] \tag{2.8}$$

From this equation:

$$C_{(x,t)} = \left\{\left[(C^o - C_o)/2\right]\mathrm{erfc}\left[\left(nx - vt/2(2Dnt)^{1/2}\right)\right]\right\} + C_o \tag{2.9}$$

We permit

$$\left\{\left[(C^o - C_o)/2\right]\mathrm{erfc}\left[\left(nx - vt/\left(2(2Dnt)^{1/2}\right)\right)\right]\right\} = A \tag{2.9a}$$

In this case, Eq. 2.9 becomes:

$$C_{(x,t)} = A + C_o \tag{2.10}$$

In Eq. 2.10 at $t = $ constant > 0 and when $x \rightarrow \infty$ parameter $A \rightarrow$ min and at very small values, the following is true:

$$C_{(x,t)} \approx C_o \tag{2.11}$$

Fig. 2.3 Graphical concept of the PM as $l_k = x(C, t)$

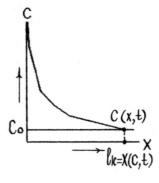

Equation 2.11 suggests that (Fig. 2.3)

(a) the migration of the pollutant with initial C^o is finite and its future penetration in the rock (matrix) does not change existing geochemical conditions;
(b) the calculation of the PM or l_k is true only at

$$C_{(x,t)} > C_o;$$

According to Eqs. 2.10 and 2.11, for a pollutant (or chemical element), the physical significance of the PM is the following. When the conditions take place:

$$\left(\partial \widehat{C}/\partial t\right) = C_o$$

The function $C = f(x,t)$ is intersected with $C_o = $ constant in the point $C(x_i, t_i) = C_o$. The graphical coordinates of the PM are $l_k = x(C,t)$ (see Fig. 2.3).

In Eq. 2.9a, only parameter x is inconstant at $x \to$ max the value of $A \to$ min. If $A = 0$ the equation $C(x_i, t_i) = C_o$ is true. The difference $(C^o - C_o)$ influences the value of A. First of all, value A is connected with the analytical determination accuracy of the studied chemical element. For example, the analytical accuracy for chloride (Cl) is 3 mg/l. Therefore, Eq. 2.11 for chloride is true only at $A \le 3$. For other chemical elements, the parameter A has its own analytical accuracy.

From a statistical point of view, the PM is a complex function as follows:

$$l_k = f(v, t, D, G) \tag{2.12}$$

where

G gradient of the chemical concentration (or simply, $G = C^o - C_o$)

It is necessary to evaluate the significance of each parameter in Eq. 2.12. Likewise, we allow the following theoretical conditions and data (as an example):

(a) The studied element is chloride, and its migration is in accordance with Eq. 2.10.
(b) The PM or l_k is calculated with $A \le 3$
(c) Initial data: $D = 0.00121$ m^2/day

$$n = 0.5$$
$$v = 0.005 \, \text{m}^3/\text{day}$$
$$t = 90 \, \text{days}$$
$$C^o = 60 \, \text{mg}/\text{l}$$
$$C_o = 10 \, \text{mg}/\text{l}$$

The indicated data are changed in sequence one by one. Graphical results of this sensibility analysis are presented in Fig. 2.4. Functional relationships are as follows:

$$l_k = 180v + 1.1 \tag{2.12.1}$$

$$l_k = 0.01t + 1.1 \tag{2.12.2}$$

$$l_k = 377D + 1.6 \tag{2.12.3}$$

$$l_k = -1.807 \ln(n) + 0.78 \tag{2.12.4}$$

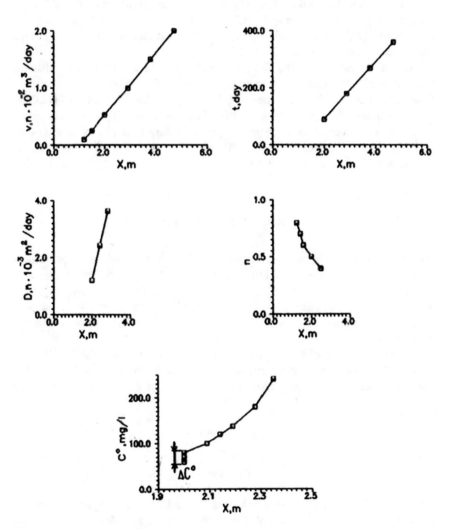

Fig. 2.4 Relationship of the PM from v, t, D, n, and C^o

Table 2.1 Results of computing values of Δl_k, (m)

PM	Parameters				
	v	t	D	C^o	n
Δl_k	0.9	0.9	0.48	0.22	−1.25

$$l_k = 0.381 \ln C^o + 0.613 \qquad\qquad (2.12.5)$$

It should be noted especially that relationship $l_k = f(C^o)$ has the following properties. A small interval of the variation C^o as ΔC^o exists in the beginning of the graphic (see Fig. 2.4). The ΔC^o value of the l_k is practically not changed for the given initial data, ΔC^o = 60–80 mg/l Cl or 33% (20 mg/l) from the initial C^o. After the interval ΔC^o, the function $l_k = f(C^o)$ is characterized by Eq. 2.12.5. Parameters v, T, D, n, and C^o were changed twice compared with initial theoretical data, and the Δl_k is the l_k for this example (Table 2.1).

Information from Fig. 2.4, Table 2.1, and Eqs. 2.12.1–2.12.5 shows that the increase of v, t, D, and C^o leads to an increase of l_k. Only increasing n reduces l_k. Parameters v and t influence l_k the most. Changes of C^o in the source of pollution have a minimal influence on the total value of the PM. In our hypothetical example, a change in C^o from 60 to 600 mg/l (10 times) increases the PM by only 0.64 m.

Thus, numerical calculation of the PM is possible using Eq. 2.10. Parameters v, t, D, n, and C^o should be predefined.

2.4   Laboratory Determination of the PM

Calculations are similar to the previous procedure, but the migration parameters (mainly D and n) are determined in laboratory conditions. Firstly, special samples should need to be collected directly in field conditions. Secondly, migration parameters will be determined in laboratory conditions by using special methodologies and equipment. Rosali (1980) proposed techniques which can be summarized as follows.

A column of rock of the length l and area S is loaded with NaCl solution at constant flow discharge q, chloride content C^o, and concentration in the rocks C_o. Chloride content is constantly recorded at the outlet of the column. Under these conditions:

$$C(x, o) = C^o$$
$$C(o, t) = C^o$$
$$C(\infty, t) = C_o$$

The solution of Eq. 2.3 is similar to Eqs. 2.4 and 2.5. These equations serve as the basis for processing the concentration–time curves of the chloride concentration in coordinates $(\xi\, t^{1/2}) - (t)$. Migration parameters are determined from this graph by the formula:

$$\xi = (ql/4S)\left\{t^{o}/\xi^{o})^{2}\right\} \text{ and } n = q\,t^{o}/Sl \qquad (2.13)$$

where

t^{o} and ξ^{o}—values estimated on the graph $\left(\xi\, t^{1/2}\right) = f(t)$ at the intersection of the straight line with axes (i.e., they are numerically equal to the lengths of the respective sections).

Rosali (1980) proposed the calculation of relative errors for n as δ_{n}, for D as δ_{D}, required column length at δ_{n}, and the time t_{E} at which δ_{n} and δ_{D} will not exceed the admissible values. The examples of the calculation of migration parameters are given in Table 2.2.

2.5   Statistical Determination of the PM

Values of longitudinal dispersion and effective porosity in the unsaturated zone are highly variable even at small intervals (e.g., data from Table 2.2). Likewise, the distribution of chemical element contents is variable in the unsaturated zone. As a rule, relationship $C = f(x)$ (x is depth, m) is a complex function, which is difficult to describe statistically. Let us analyze this function in detail. Data for D and n (using chloride as the pollutant) are summarized in Table 2.3.

Calculation of $C(x,t)$ is achieved using data from Table 2.3, and values of v, t, C^{o}, and C_{o} are required for solving Eqs. 2.12.1–2.12.5. The value of PM is $l_{k} = 3.0$ m for $C_{o} = 10$ mg/l (Fig. 2.5).

From a theoretical point of view, and according to experimental data (e.g., determination of migration parameters), movement of a chemical element in the unsaturated zone is accumulative. In other words, chemical elements are transported through the unsaturated zone by the piston flow model. The theory is described in many publications. Shestakov (1973) and Dagan (1987) published an analysis of the general theory of the piston model. Ellsworth et al. (1991) published a practical application of

Table 2.2 Data related to migration parameters of the unsaturated zone, Republic of Moldova

Well number (m)	Sampling depth (m)	k (m/day)	q (m/day)	v (m/day)	D (m²/day)	δ_{D}	n	δ_{n}	l (m)	l_{E} (m)	t (day)	t_{E} (day)
2	2.1	0.047	0.12	0.30	0.009	−0.1	0.88	0.16	0.17	0.18	1.25	1.23
3	0.4	0.01	0.036	0.09	0.001	0.0	0.85	0.08	0.17	0.18	3.40	3.23
11	4.0	0.06	0.10	0.25	0.003	0.0	0.7	0.07	0.17	0.20	1.11	0.94
18	9.0	0.018	0.034	0.084	0.028	−0.1	0.37	0.17	0.17	0.19	2.04	2.00

Remark: k is hydraulic conductivity; other parameters are defined in text

Table 2.3 Values of longitudinal dispersion (D) and effective porosity (n) for sandy clay, Republic of Moldova

Number of the sample	Depth of sampling (m)	D (10^{-3} m²/day)	n
1	1.0	0.8	0.5
2	1.5	12.0	0.6
3	2	1.0	0.4
4	2.5	40.0	0.7
5	3.0	4.0	0.65

Fig. 2.5 Steady-state distribution of chloride

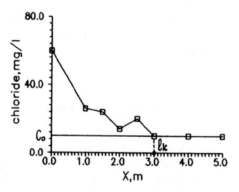

the flow piston model using three inorganic tracers: Ca(NO₃)₂, KCl, and H₃BO₃. In this context, we propose to calculate the cumulative concentrations of chloride from Table 2.3 and Fig. 2.5 in the following way (based on the piston model theory):

$$x = 0.0 \quad C = 60 \text{ mg/l}$$
$$x_1 = 1.0 \quad C^1 = C + C_1 = 88 \text{ mg/l}$$
$$x_2 = 1.5 \quad C^2 = C^1 + C_2 = 113 \text{ mg/l}$$
$$x_i = m \quad C^n = C^{(n-1)} + Cn$$

where

$$i = 1, 2, 3 \ldots \text{ and } m > 0.$$

Fig. 2.6 Cumulative concentration of chloride

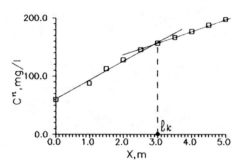

The new graphic, created with the cumulative function $C^n = f(x)$, is shown in Fig. 2.6. Analysis of Fig. 2.6 indicates that the cumulative concentration of chloride consists of two simple linear functions. Its intersection of the x-axis is the value of the PM, or $l_k = 3.0$ m in our example. The results of the PM determination are the same for calculation using Eq. 2.10 (Fig. 2.3) and the statistical cumulative method (Fig. 2.6).

The relationship $C = f(x)$ in natural conditions has a complex distribution and is often difficult to describe mathematically. It is not possible to determine the PM directly from such signals. This is only possible using the new transformed cumulative function: $C^n = f(x)$. Figure 2.7 confirms this.

Geochemical cumulative properties are traced for Cl^-, HCO^{3-}, SO^{4+}, K^+, Ca^{2+}, Mg^{2+}, Na^+, NO^{3-}, Cu^{2+}, and TDS. It is very probable that these cumulative properties are characteristic of a large spectrum of chemical elements.

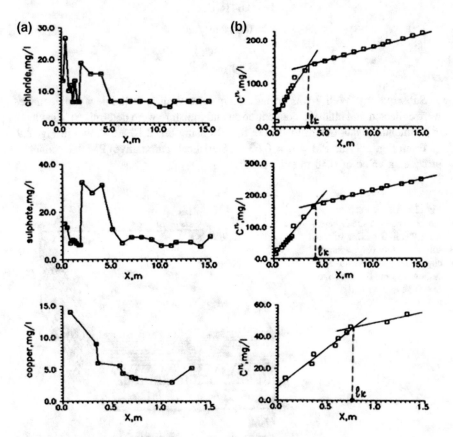

Fig. 2.7 Examples of chemical element distributions in the unsaturated zone, Republic of Moldova (where **a** represents steady-state signals and **b** represents cumulative transformation)

2.6   Experimental Determination of the PM

In the territories of Moldova, the USA, and Germany, experimental determinations of the PM are carried out in natural conditions. The Moldova site is representative of our study. The block—scheme of the test site—is displayed in Fig. 2.8.

The methodology of the experiment was simple. Samples from the unsaturated zone were collected in Well 3 up to a depth of 1.2 m with an interval of 10 cm. The conditions of background chloride concentration, total moisture, and migration parameters D and n were determined in the laboratory.

The solution of NaCl infiltrated the zone of aeration through a metallic ring with diameter 0.22 m and height 0.245 m. The zone was characterized by the following data:

$$t = 0.0416 \text{ day}$$
$$v = 0.504 \text{ m}^3/\text{day}$$
$$C^\circ = 330 \text{ mg/l}$$
$$C_o = 19 \text{ mg/l}$$

Subsequently, Well 2 was drilled 25 cm from Well 1. In the new well, samples were collected and chloride concentration and moisture were determined. Results of experimental in situ determination of the PM using Eq. 2.10 are shown in Fig. 2.9.

From Fig. 2.9, the PM = l_k = 0.63 m. Statistical (cumulative) PM determination provides a value of 0.58 m (Fig. 2.10).

Fig. 2.8 Block—scheme of the experimental site (where *1* well; *2* soil; *3* mixing of the soil and sandy clay; *4* sandy clay; *5* sandy clay with carbonates; and *6* grass cover) $D = 0.32$ m^2/day and $n = 0.40$

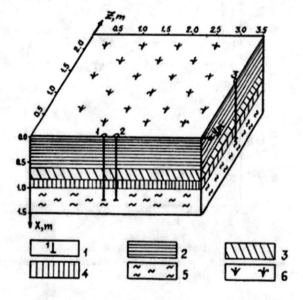

Fig. 2.9 Experimental and
numerical determination of
chloride concentration and the
PM (where *1* experimental
and *2* numerical)

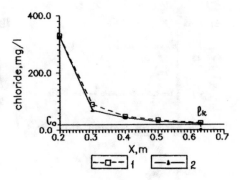

Fig. 2.10 Statistical PM
determination

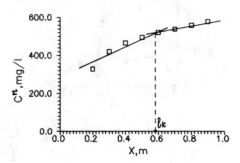

Fig. 2.11 Distribution of the
moisture in the unsaturated
zone (where *1* moisture in the
well after experiment, *2*
moisture in natural
conditions)

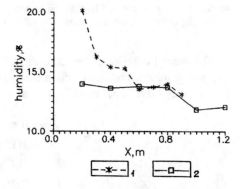

Analysis of the experimental, numerical, and statistical PM determinations
suggests that values of (l_k) are comparable. A small deviation of 0.05 m is
explained by the error of the statistical method.

The actual PM is verified by the moisture data of samples from Wells 2 and 3
(Fig. 2.11). As it is evident from this graph, moisture lines at depth 0.60 m coin-
cide. These indicate that chloride migrated together with water flow up to this
depth.

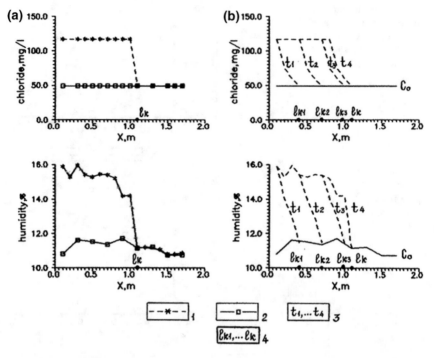

Fig. 2.12 Graphical results of the field experiment (where *1* numerical determination of chloride concentration and moisture, *2* background (natural)values of chloride concentration and moisture, *3* time (days), and *4* movement (formation) of the PM)

The second experiment was also carried out in Moldova, albeit in other geological conditions. The cross section contains the intervals 0.0–0.6 m soil and 0.6–1.7 m sandy clay. Initial conditions were as follows:

$$v = 1.0 \text{ m/day}$$
$$t = 0.0416 \text{ day}$$
$$C^o = 117 \text{ mg/l}$$
$$C_o = 49.0 \text{ mg/l}$$

The results of this experiment are displayed graphically in Fig. 2.12.

Statistical determination of the PM is displayed in Fig. 2.13. Analysis of the data from Figs. 2.12 and 2.13 confirms that PM = l_k = 1.30 m, the same for both experimental and statistical data.

Fig. 2.13 Statistical
determination of the PM for
experimental data

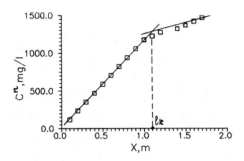

2.7 Geochemical Aquifer Vulnerability Estimation Leakage Potential Methodology (GAVEL)

In previous chapters, the PM was determined with numerical, laboratory, statistical, and experimental methods. It was also demonstrated that statistical methodology is the most convenient from a practical and cost-effective point of view. It is necessary to include the PM in the procedure for vulnerability estimation of the groundwater. In this context, the following assumptions are considered:

1. Vulnerability is estimated for an unconfined aquifer system, which is not directly exposed to land surface, but is overlaid by the unsaturated (or aeration) zone.
2. The PM should be understood as the determination of multiannual maximum depth of pollutant penetration in the different types of rocks and soil.
3. Theoretically, the PM will be moving slowly in the unsaturated zone if the thickness of infiltrated water H_w is more than zero ($H_w > 0.0$) and H_w = constant in time. Such cases are characteristic of ponds, artificial lakes, etc.
4. The PM is an integral parameter involving the variation of hydrodynamic longitudinal and transversal dispersion as well as the geochemical transformation of a pollutant during migration.
5. When PM or l_k is equal to the water table value H ($l_k = H$), the existence of pollution of the aquifer is de facto existing and determination of the PM does not make sense.

Geochemical Aquifer Vulnerability Estimation Leakage potential method (GAVEL) is a simple model expressed by the formula:

$$R = l_k/H_s$$

where

R geochemical aquifer vulnerability estimation leakage potential,
l_k point of pollutant migration in the unsaturated zone (PM, m),
H_s static aquifer water table, m.

In physical terms, GAVEL estimates the risk of the probable or existing groundwater pollution. This method can be used for:

(a) point vulnerability estimation,
(b) regional survey, and
(c) mapping of the vulnerability on different topographical scales.

The following classification of the vulnerability of groundwater is proposed:

1. $R < 0.25$—risk of the groundwater pollution does not exist or it is low (LR)
2. $0.25 < R < 0.5$—medium risk of the groundwater pollution (MR)
3. $0.5 < R < 0.75$—high risk of the groundwater pollution (HR)
4. $R > 0.75$—very high risk of the groundwater pollution or already polluted (VHR).

The flowchart of the practical application of the GAVEL is presented in Fig. 2.14.

2.8 Discussion

It is evident and indisputable that existing methods of groundwater vulnerability assessment cannot be applied internationally. Perhaps, this is because the used methodologies are imperfect, or collection and input of the special required data are problematic. At the same time, modern methods of groundwater vulnerability assessment are connected with the problem of spreading pollution in the unsaturated zone. The worldwide interest of many researchers, environmentalists, and engineers is leaning in this direction. Penetration of harmful pesticides in water-bearing horizons and health-related consequences has amplified the practical and theoretical interests of improving the methodology of groundwater vulnerability approaches. Climate change or global warming is also in this target. It is crucial in the vulnerability assessment to understand the migration of pollutants in the unsaturated (or vadose, or aeration) zone.

In this way, we propose a geochemical method based on the Geochemical Aquifer Vulnerability Estimation Leakage Potential (GAVEL), with the goal of description and argumentation of the pollutant spread using simple geochemical vertical signals. In our opinion, this is a cheaper and more efficient way of approaching the vulnerability problem.

The simple and logical scheme of pollution of an unconfined aquifer consists of three elements:

Depth of sampling, m	Step 1 - Original data	Step 2 - Cumulative data
1	5	5
2	10	15
3	3	18
4	15	33
5	8	41
6	4	45
7	21	66
8	6	72
9	17	89
10	3	92
11	6	98
12	2	100
13	4	104

Step 3 - Cumulative graphic and graphical determination of the l_k

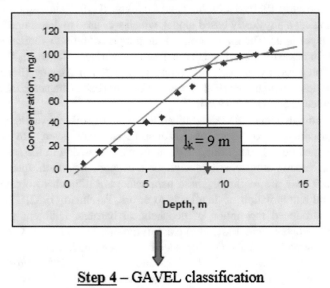

Step 4 – GAVEL classification

$$R = l_k / H_s$$

Fig. 2.14 Methodological flowchart of the GAVEL

1. Surface sources of pollution.
2. Migration of the pollutant in the unsaturated zone.
3. Infiltration of the aquifer by pollution.

In many cases, the unsaturated zone combines the soil stratum and sequences of different strata of parental rocks. As a rule, the thickness of the zone is well known. Numerically, it is the distance between the land surface and the water table (strictly speaking, it is the statistically averaged water level due to seasonal or anthropogenic fluctuations). A water-bearing horizon (aquifer), like an unconfined one, is characterized by geometrical and hydrodynamic parameters, which are mostly determined. The surface source of pollution (e.g., point type) is active during a period of time, which is known or can be measured. If the thickness of the unsaturated zone is big enough (in practice > 5 m) and not homogeneous, it is probable that the pollutant does not reach the aquifer. It only migrates to one point (position) in the range of the unsaturated zone. We proposed this point as the **P**oint of pollutant **M**igration (PM) and denote it as (l_k). If the value (l_k) is known, we can argue prognoses about groundwater quality for point and nonpoint sources of pollutants. Typically, the parameter (l_k) is characteristic of irrigation, landfills, infiltration of contaminated atmospheric precipitations, etc.

The concept (l_k) is described insufficiently in scientific literature. Voss (2003) selected PRZM-2 model for simulating pesticide leaching because this method has been used and tested extensively in many studies since its development by the USEPA in the early 1980s. Holtschlag and Luukkonen (1997) investigated leaching of atrazine using a physically based model, which was used to determine the time of travel of the pesticide. The main formula is an integral of ratio to vertical velocity of water with indicated depth of interest below land surface. Numerous publications have been prepared by the US Environmental Protection Agency (Lead leaching 1994; A review of methods 1993; Factor affecting 1988; Characterization of soil 1984; Polluted groundwater 1973).

Polubarinova-Kocina (1952)determined the depth of soil moistening (y_0), which can be approximately considered as (l_k) for soil strata only. Avereanov (1965) also proposed a method for computing the depth of water leaching in the soil zone. For this method, precise atmospheric precipitation, evapotranspiration, meteorological, and soil-rock data are required. These parameters are difficult to obtain for each territory, and extrapolation of data leads to errors. Perelman (1982, 1989) for the first time introduced the notion of geochemical barriers. Different geochemical barriers (oxygenated, acid–base, etc.) exist in the unsaturated zone. Chemical elements can migrate until encountering specific barriers, which are characteristic of selected geochemical associates. Wellings and Bell (1980, 1982) proposed ideas whether are much closed to GAVEL. These authors describe the physical controls of water movement in the unsaturated zone. They propose the concept of the zero flux plane (ZFP). This is the plane in the profile where the total potential gradient is zero, and it divides the profile into zones of upward and downward flux. The ZEP is controlled by seasonal fluctuation of atmospheric precipitation and hydraulic conductivity of porous media. Gurdak (2008) is trying to use the ZEP, but only in logistic context.

The proposed GAVEL methodology with PM determination has no analogues in the official published scientific literature. Analysis of the methods for PM determination reveals that migration of the chemical pollutants in the unsaturated zone at $C^o > C_o$ depends on the time of filtration. The nature of this is explained by the dispersion and piston flow models. Full dispersion of the pollutant takes place at the point (l_k), and in the point $x = l_k$, dispersion is equal to zero. At $x > l_k$, migration of the pollutant does not take place. Analogous processes are characteristic of the distribution of the heat in solid materials (Carlslaw and Jaegar 1959).

Results of the numerical, laboratory, statistical, and experimental methods are practically the same and compatible. The most interesting and practical is the statistical method, which has the following qualities. Dividing of the cumulative function $C^n = f(x)$ into two linear functions can probably be explained as a geochemical process. In natural conditions (no atmospheric precipitation, irrigation, and other anthropogenic impact), cumulative distribution of the chemical elements is linear (at least should be theoretically). Under the influence of infiltration of different kinds of water on the top of the unsaturated zone, geochemical conditions and equilibrium are changed. This influences the redistribution of the chemical elements in the frame of the unsaturated zone. Such a process is depicted in the cumulative function as consisting of two linear ones.

The evidence of the geochemical cumulative process is the experimental data, which are described in this investigation. The representative example is shown in Fig. 2.15. Natural distribution of both the chloride and humidity in a cumulative sense is a single linear function. Distribution of the same parameters after infiltration of the solution NaCl is drastically changed. In this case, the cumulative

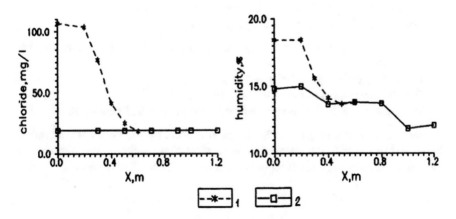

Fig. 2.15 Experimental distribution of chloride and humidity in the unsaturated zone (where *1* natural distribution and *2* distribution after infiltration of the solution NaCl)

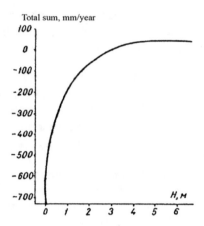

Fig. 2.16 Total sum of water balance versus thickness of the unsaturated zone (total sum = évapotranspiration − infiltration, H–depth)

function as a rule consists of two intersected linear functions. Projection of the intersection point on the x-axis (depth, m) is the numerical value of the PM or (l_k).

The PM is strongly connected with hydrogeological conditions and particularly with the process of groundwater recharge from atmospheric precipitation. A multiyear statistical balance is displayed for the Republic of Moldova in Fig. 2.16 (Zelenin 1984). Wellings and Bell (1980, 1982) describe the same idea.

From this graph (see Fig. 2.16), it is evident that equality between evaporation and infiltration is located at a depth of approximately 3.0 m in the unsaturated zone. The average PM in Moldova is also around 3.0 m. Additional confirmation of this fact is illustrated in Fig. 2.17 (Zelenin 1984). The graphical data suggest that water balance parameters are more expressive at the depth of the unsaturated zone equal to approximately 3.0 m.

Thus, the PM determination is more practical in the case of application of the statistical method. The main advantages are as follows:

1. It is not necessary to determine migration parameters in both laboratory and field conditions,
2. The results are adequate for natural conditions,
3. The PM determination is not time-consuming,
4. Statistical determination does not require highly qualified personnel.

Generally, the PM or l_k has a significant theoretical hydrogeological importance, because if this is true, the classical point of view about recharge and discharge of the unconfined aquifers needs to be revised and conceptually corrected.

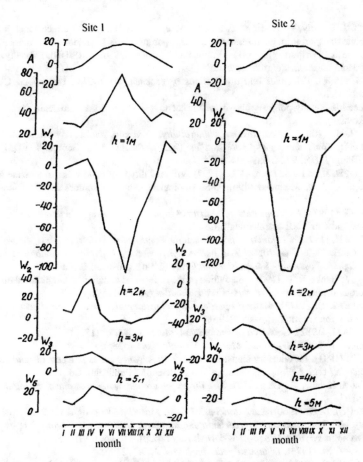

Fig. 2.17 Statistical multiyear changes of the atmospheric temperature (T, °C), monthly sum of precipitation (A, mm), and total sum W = evapotranspiration − infiltration (mm) for two sites, South Moldova (where h is water table, and W1 − W5 are different values of total sum of water balance)

References

Avereanov, S. F. (1965). Some question related to salinisation prognosis of the irrigated land and methods of contestation in the European part of USSR. In *Irrigated lands in the European parts of USSR* (174p), Moscow, Kolos.

Bhadoria V. 2013. *Introduction to some simple signal.* http://www.scribd.com/doc/163208161/Introduction-to-Some-Simple-Signal.

Carlslaw, H. S., & Jaegar, J. C. (1959). *Conduction of heat in solids* (510p). Oxford: Oxford University Press.

Cook, P. G., Edmunts, W. M., & Gaye, C. B. (1992). Estimating paleorecharge and paleoclimate from unsaturated zone profiles. *Water Resource Research, 10,* 2721–2731.

Dagan, G. (1987). Theory of solute transport by groundwater. *Annual Review of Fluid Mechanics, 19,* 183–215.

Edmunts, W. M., & Tyler, S. W. (2002). Unsaturated zones as archive of past climates: Towards a new proxy for continental regions. *Hydrogeology Journal, 10,* 216–228.

Ellsworth, T. R., Jury, W. A., Ernst, F. F., & Shouse, P. J. (1991). A three-dimensional field study of solute transport through unsaturated, layered, porous media 1. Methodology, mass recovery, and mean transport (p. 497). Publications from USDA–ARS/UNL Faculty. http://digitalcommons.unl.edu/usdaarsfacpub/497.

Faure, G. (1998). *Principles and applications of geochemistry* (600p). Englewood Cliffs NJ: Prentice Hall.

Geohimiceskie metody poiskov rundyh mestorojdenii. (1982). Sibirskoe otdelenie, Nedra (167p). (in Russian).

Gurdak, J. J. (2008). *Ground-water vulnerability: Nonpoint-source contamination, climate variability, and the high plains aquifer* (223p). Saarbrucken, Germany: VDM Verlag Publishing. ISBN: 978-3-639-09427-5.

Holtschlag, D. J., & Luukkonen, C. L. (1997). Vulnerability of ground water to atrazine leaching in Kent County, Michigan (49p). U.S. Geological Water—resources investigations report 96-4198.

Kay, M. S. (1993). *Fundamentals of statistical signals processing estimation theory* (589p). Englewood Cliffs NJ: Prentice Hall Inc.

Krainov, S. R. (1973). *Geohimia redkih elementov v podzemnyh vodah. Moscow, Nedra* (270p).

Moraru, C. E. (2009). *Gidrogeohimia podzemnyh vod zony activnogo vodoobmena krainego Iugo-Zapada Vostocno - Evropeiskoi platformy* (210p). Chisinau: Elena V.I.

Ogata, A., & Banks, R. B. (1961). A solution of the differential equation of longitudinal dispersion in porous media. *Geological survey professional paper* (441-A, pp. 1–7).

Pashkovskii, I. S. (1973). *Metody opredelenia infilitrationnogo pitania po rascetam vlagoperenosa v zone aeratii. Moskva* (219p).

Perelman, A. I. (1982). *Geohimia prirodnyh vod, Moscow, Nauka* (152p).

Perelman, A. I. (1989). *Geohimia. Moscow, Vyshaia skola* (528p).

Priemer, R. (1991). Introductory signal processing. In *Advance series in electrical and computer engineering* (vol. 6, 730p). Singapore: World Scientific Publishing Co. Pte.

Polubarinova-Kocina, P. Ya. (1952). *The theory of ground water movement. Moscow, Nauka* (380p).

Radushkevich, L. B. (1960). *Kurs staticheskoi fiziki.* Moscow: Uchpedgiz. (in Russian).

Rosali, A. A. (1980). *Metody opredelenia migrationnyh parametrov. Obzor VNII ekonomike minera'nogo syrya i geologo-razvedocnyh rabot* (62p). Moscow: VIEMS.

Shestakov, V. M. (1973). *Dinamika podzemnyh vod. MGU* (210p).

Taylor, J. (1954). The dispersion of mater in turbulent flow through a pipe. *Proceedings of the Royal Society, Series A, Mathematical and Physical Sciences, 223*(1155), 446–468.

U.S. EPA. (1993). *A review of methods for assessing aquifer sensitivity and ground water vulnerability to pesticide contamination.* EPA# 813R93002.

U.S. EPA. (1984). *Characterization of soil disposal system leachates.* EPA# 600284101.

U.S. EPA. (1988). *Factors affecting trace metal mobility in subsurface soils.* EPA# 600288036.

U.S. EPA. (1994). *Lead leaching from submersible well pumps [environmental fact sheet.* EPA# 747F94001.

U.S. EPA. (1973). *Polluted groundwater: Some causes, effects, controls, and monitoring.* EPA# 600473001b.

Voss, F. D. (2003). *Development and testing of methods for assessing and mapping agricultural areas susceptible to atrazine leaching in the State of Washington* (13p). U.S. geological survey water—Resources investigation report 03-4173.

Wellings, S. R., & Bell, J. P. (1980). Movement of water and nitrates in the unsaturated zone of upper chalk near Winchester, Hants, England. *Journal of Hydrology, 48,* 119–136.

Wellings, S. R., & Bell, J. P. (1982). Physical control of water movement in the unsaturated zone. *The Quarterly Journal of Engineering Geology, 15,* 235–241.

Zelenin, I. V., et al. (1984). *Interrelation of ground and surface water of Moldova. Chisinau, Stiinta* (150p).

Chapter 3
Basic Principles of Hydrogeology for Hydrogeochemical Vulnerability

Constantin Moraru and Robyn Hannigan

Abstract Hydrogeochemical vulnerability is a part of groundwater vulnerability study. At present time, different meaning of pollutant, unsaturated zone, and aquifer is proposed in scientific literature. The analysis of these basic terms has been done. Contamination, pollution, and pollutants are differentiated. Unsaturated zone or vadose zone notion is analyzed related to the groundwater vulnerability. The modern meaning of an aquifer is described from the view of water quality protection.

Keywords Water contamination and pollution · Unsaturated or vadose zone
Aquifer

Hydrogeochemical vulnerability is defined as the natural properties of an aquifer and surrounding geological environment to support the anthropogenic pressure, particularly from pollutants, on water quality. When studying hydrogeology, it is important to include hydrogeochemistry or groundwater geochemistry as subdiscipline that should be discussed separately. Many publications describe groundwater vulnerability; however, there are varying levels of understanding of hydrogeological principles. This chapter will focus on and describe three interconnected logical blocks associated with hydrogeochemical vulnerability: pollutants, unsaturated zones, and aquifers.

3.1 Pollution and Pollutants

First, we need to differentiate contamination and pollution of groundwater. In the literature, they are considered synonyms with the same meaning (Fetter 2001); only pollution is used (Todd 1980); or only contamination is described (Hudak 2000). Secondly, sometimes both pollution and contamination are used interchangeably without a clear understanding (Soliman et al. 1997). Freeze and Cherry (1979) proposed that all solutes introduced in the hydrologic environment because of man's activities are referred to as contaminants, regardless of whether or not the

© Springer International Publishing AG 2018

C. Moraru and R. Hannigan, *Analysis of Hydrogeochemical Vulnerability*,
Springer Hydrogeology, https://doi.org/10.1007/978-3-319-70960-4_3

concentrations reach levels that cause significant degradation of water quality. This author then reserves the term pollution for situations where contaminate concentrations attain levels that are considered to be objectionable. Shestakov and Pozdneakov (2003) suggested that pollution is a negative effect of groundwater quality changes and contamination is the process, which the pollution is formed.

On the one hand, we know that the French origin of the word contamination means water quality is changed from natural state to varying levels of physical, chemical, and microbial parameters. For example, we have the following initial conditions: monitoring well and time series for chemical water composition, and the World Health Organization (WHO) hygienic standards are used for drinking water. Out of all of our parameters, we selected nitrates (NO_3) with the WHO standard equal to 50 mg/l. Over a certain period, there was a statistical trend where nitrate concentration ranged between 10 and 20 mg/L. That period in time (t), when the NO_3 concentration starts to increase with fluctuation to 25, 30, and up to 50 mg/L over a slow period, is an example of *contamination*. In other words, contamination is a process where concentration changes in groundwater within the limits of an official water quality standard. Because of varying standard values, the limits of contamination can also vary. Groundwater *pollution* occurs when the concentration of standardized parameters goes above the suggested limits. In the case with NO_3, the pollution that happened after the concentration of nitrates rose above 50 mg/l. Thus, contamination and pollution are strictly related to existing water quality standards. There are however parameters, such as sodium and carbonate, that are naturally high in groundwater; therefore, such ecological limits do not exist.

On the other hand, pollutants come from different sources and can belong to chemical, microbiological, radioactive, and thermal categories. Todd (1980) describes in detail the various sources of pollutant for groundwater. He discussed municipal, industrial, agricultural, and miscellaneous sources and causes as well as analyzed and evaluated pollution potential (Todd 1980). According to the chart of evaluation, a rating table is proposed and it takes into account depth to water, sorption above water table, permeability, water table gradient, and horizontal distance of the source (Table 3.1). Using this, the method permits the optimization of research and applied work related to groundwater vulnerability estimation.

Committee on Techniques (1993) states a fundamental principle: *All groundwater is vulnerable*. There are different classifications of pollution sources, and most of them are conventional and without practical importance. Nevertheless, sources of pollution are divided into two groups. The first group is known as point

Table 3.1 Possibility of groundwater pollution (Todd 1980)

N/n	Total point	Possibility of pollution
1	0–4	Imminent
2	4–8	Probable or possible
3	8–12	Possible but not likely
4	12–25	Very improbable
5	25–35	Impossible

sources, and the second one is known as diffuse (or non-point) sources of pollution. These groups are not defined by numerical values or connotations. Often point sources become diffuse sources and vice versa. A special case arises where, for example, a power station may emit sulfur dioxide and nitrous oxide to the air. Although this is a *point source*, the deposition (fallout) and, hence, impact will be over a wide area as *diffuse pollution* (Sources of pollution 2014). Examples of point sources are described in many European countries (COST action 620 2003) and non-point sources for High Plains Aquifers, USA (Gurdak 2008).

It is more practical and logical to specify the sources of pollution according to their origin: chemical (organic and inorganic), microbiological, radioactive, thermal, etc. This allows groundwater vulnerability estimation to be done methodologically, producing important and reputable research.

3.2 Unsaturated Zone

It is widely known that unsaturated zone or vadose zone is synonymous. The term unsaturated zone is mostly used by geologists and hydrogeologists, without taking the origin of soil cover and parental rocks into consideration. Vadose zone is described by both the geologists and the soil and engineering specialists with the focus on the upper part. In our case, we will consider both unsaturated and vadose zone as the same hydrogeological structure. This is because this zone is between land surface and the first aquifer, typically unconfined. Unfortunately, the unsaturated zone is not studied in detail. Soil specialists are involved in the study of the soil strata, which is important for agriculture. For geologists, this zone does not contain mineral deposits causing geological data about the unsaturated zone to be limited. For instance, in the deep borehole logs, the upper part (unsaturated zone) is described and referred to as only as a small part of Quaternary deposits.

There are many definitions for the unsaturated zone. We support Wight and Sondereger (2001) who state that unsaturated zone is composed of the materials from the land surface down to the water table, including the capillarity fringe. Thickness of the unsaturated zone is from 0.5 m (Moldova) until hundreds of meters in the Netherlands. It is considered that thickness of the zone is not changed, and it is static in time. This concept is used in many applied models. Recent investigations demonstrate that thickness of the unsaturated zone changes with time (Moraru and Anderson 2005; Gurdak 2008; Moraru 2009). Therefore, variations of the water table of an unconfined aquifer imply changes in thickness of the unsaturated zone (Fig. 3.1).

In some cases, groundwater level can vary from 0.5 to 1.5 m (Moraru 2009). This indicates that the thickness of the unsaturated zone also changes in the indicated intervals. This fact is important and needs to be considered before groundwater vulnerability is measured. Such data are of particularly higher importance for prognostic models and decisions.

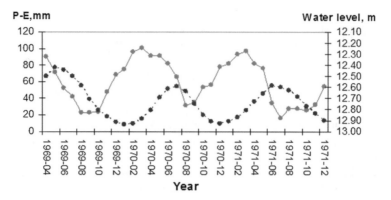

Fig. 3.1 Relationship between atmospheric precipitation-evaporation (P-E) and water level in selected well Sh:P-99, Memphis (USA) (continuous line—difference precipitation-evaporation; discontinuous line–water level in the monitoring well) (Moraru and Anderson 2005)

Water flow in the unsaturated zone is different compared to the flow in water-saturated media (aquifers). The continuity equation, while commonly used to simulate unsaturated flow and transport, does not consider the validity of Darcian flow (Tindal et al. 1999). Hydraulic conductivity of the zone is also different when compared to saturated media. This parameter depends on the initial humidity of the zone and the infiltration potential. Usually, potential is low and depends on the volume of atmospheric precipitation. Factors affecting unsaturated hydraulic conductivity and development of unsaturated flow equation are well described (Tindal et al. 1999; Wight and Sondereger 2001).

Infiltration of water into zone depends on the angle of decline at the site, vegetation cover, and lithological properties. Shestakov and Pozdneakov (2003) studied this question in detail. Infiltration of water in the unsaturated zone is a concept that is understood to be a component of recharge for an unconfined aquifer. Selected and resent research clearly indicates that water moving in the frame of the zone is limited by distance. Wellings and Bell (1980, 1982) describe the physical controls of water movement in the unsaturated zone. They propose the concept known as the zero flux plane (ZFP). This is the plane in the profile where the total potential gradient is zero, and it divides the profile into zones of upward and downward flux. The ZEP is controlled by seasonal fluctuation of atmospheric precipitation and hydraulic conductivity of porous media. Moraru (2009) and the present publication argued that water as "carrier" for pollutants also have a final point of infiltration in the unsaturated zone.

Natural and anthropogenic processes influence the geochemistry of the unsaturated zone. Many publications are committed to answer this question (Faure 1998; Tindal et al. 1999; Moraru 2009). It is important to mention that from a geochemical point of view, the unsaturated zone is not homogenous. Perelman (1989, 1982) introduces a concept known as the geochemical barrier. Barrier means that chemical elements are retained in particular levels within the unsaturated zone because of changes in geochemical conditions, like system pH-eh, concentration of O_2, and other chemical elements. For example, Cu cannot migrate to the deep areas of the zone if the conventional depth is located at $CaCO_3$ (carbonate barrier). This is because a new complex substance is formed, like $CuCO_3$. Geochemically, Cu is stopped at the barrier surface causing the unconfined aquifer to not be polluted with copper. The detailed investigation of the Chisinau City (Republic of Moldova) demonstrates that an unconfined aquifer is not polluted with trace elements, because of the effect of different geochemical barriers (Myrlean et al. 1992).

Water balance in the unsaturated zone is the main problem with different models and practical applications. Shestakov and Pozdneakov (2003) propose original methods. Tindal, Kunkel, Anderson (1999) and Wight and Sondereger (2001) describe in situ determination of the elements in the water balance. These include direct measurements of infiltration rates. Calculated water balances very often contain many mistakes. Moraru and Zincenco (2005) note that, the calculated value of the infiltration into unsaturated zone is 2.1×10^{-5} m/day for the territory of Chisinau City. MODFLOW model accepts only 0.021×10^{-5} m/day for steady-state condition of the unconfined aquifer. The difference here is about 10 times less compared with water balance calculations.

Unsaturated zones contain different gases. The transport of heat and gas in the zone is compared with water, according to other rules. Evaporation and condensation of water are fully controlled by the temperature of the upper part or soil cover. Wight and Sondereger (2001) analyzed soil through heat transfer, conductivity, and diffusivity of heat capacity. These authors determined that soil also has the ability to transport.

3.3 Aquifer

In many cases, unconfined aquifers are the target for vulnerability estimation. Analysis of the groundwater vulnerability methods showed that a spectrum of methods is designed for this purpose and mainly for shallow groundwater. Goldberg methodology (Goldberg 1987; Goldberg and Gazda 1984) is the only exception, which estimates vulnerability for both unconfined and confined aquifers (see, Chap. 1, this publication).

An aquifer is the major part of structural hydrogeology. There is not one unique definition of an aquifer. Of all the words in hydrogeology, there are probably none with more shades of meaning that the term aquifer (Freeze and Cherry 1979). An aquifer is a geological unit that can store and transmit water at rates fast enough to supply reasonable amounts to wells—mention Fetter (2001). Todd (1980) notes that a geological formation that yields significant quantities of water can be defined as an aquifer. Wight and Sondereger (2001) have other definition—an aquifer is a formation or part of a formation that contains sufficient saturated permeable material to yield significant quantities of water to wells or springs. Freeze and Cherry (1979) have a more realistic definition—an aquifer is a best-saturated permeable geological unit that can transmit significant quantities of water under an ordinary hydraulic gradient. An aquifer is a rock unit that will yield water in a usable quantity to well or spring (Heath 1987). The list of various definitions can go on, but we will not find one unique meaning. Practically, all definitions have one technical aspect, which is to provide water to well or spring. The point is that aquifer(s) are natural geological formations, and water stored in the horizon does not depend on the economical usage. What about cases where aquifers contain saline water or brines? It is clear that in the definition of the aquifer it does not necessarily include ability to yield water in economical quantities.

In order to protect groundwater, a more precise definition of aquifer is needed. In this context, we understand that an aquifer is a stratigraphic geological formation void that contains gravitational water. Such definition is similar to the same described by Freeze and Cherry (1979), Alitovskii (1962), and Polubarinova-Kocina (1952). This aquifer definition relates to the concepts of the groundwater vulnerability assessment.

Aquifers are typically classified as unconfined or confined. An unconfined aquifer, as its name suggests, is a water-bearing horizon closest to the land surface. A confined aquifer has an impermeable horizon as the top cover. There is no answer as to where the horizon is located in geological space. In many textbooks, confined aquifers are associated with the artesian type. When groundwater if flowing under hydrostatic pressure up to land surface this aquifer is known as artesian (*Artesium*, lat. Name and locality in France). Unconfined aquifers, as a rule, contain hydraulic pressure, but rarely this pressure is enough for self-flowing groundwater. When estimating vulnerability, it is more convenient to use phreatic aquifers (unconfined). In this context, deep aquifers (confined) need more hydrogeological data.

Hydrogeochemical vulnerability is the change in groundwater quality due to the influence of human activity. Modern hydrogeology does not have one internationally recognized hydrogeochemical classification. Piper diagram or classification, or Durov's, does not fully express the geochemical changes in groundwater. Moreover, each country often uses its own national classifications, for example, Russia uses Kurlov formula and Aliokin classification; USA, Piper diagram; and the Netherlands, Stuyvesant classification. The same problem is common for classifications using TDS values. The complexity of this question is well described by Moraru and Anderson (2005). In this publication, a new hydrogeochemical classification is proposed which feeds into the needs of the groundwater vulnerability assessment.

It is logical and sensible that vulnerability assessment be divided into two groups: regional and local. Regional assessments need to be correlated with the importance of an aquifer for a particular region of study. Places, which are not important for water supply or are not considered and do not need special assessment. Local vulnerability assessment is more important, and often results are used in different decisions, for models and other hydrogeological prognosis.

Groundwater geochemistry is the result of four equilibrium systems: water–rock, water–gas, water–biological parameters, and water–dissolved substances (Samarina 1977). Calculation methods of the equilibrium species are well-developed (Driver

1988; Appelo and Postma 1993; Langmir 1997, Aquachem V. 4.0 2003, Merkel and Planer–Friedrich 2005; Berkovwitz et al. 2008). One or more system is affected during the vulnerable impact on an aquifer. Often it is difficult to determine which system is equilibrated, but the leader system is appreciated. Future designs and prognoses need to be adjusted for a leader equilibrium system. Results of the equilibrium modeling and vulnerability assessment need to be correlated with the natural self-cleaning characteristic of the aquifer. This process eliminates or reduces high concentrations of a pollutant in an aquifer during the geochemical migration. Self-cleaning processes in an aquifer are occurred due to dispersion, convection, chemical transformation, and other natural and artificial mechanisms. A representative example is presented by Moraru (2009). A phreatic aquifer is strongly polluted by a livestock farm. Sources of pollution have concentrations of $NO_3 = 250$ mg/l. This is approximately 900 m from these sources, $NO_3 = 10$ mg/l and appear $NO_2 = 1.5$ mg/l. In this case, self-cleaning is due to the denitrification process.

Modern groundwater vulnerability assessment (planning, modeling, and design) need to take into account the characteristics of the sources of pollution, particularities of the unsaturated zone, and real structure of the aquifer geometry and hydrogeology. GIS technologies are a powerful instrument for this scope.

References

Alitovskii, M. E. (1962). Spravocnik gidrogeologa (p. 586). Moskva: Nedra.

Appelo, C. A. J. & Postma, D. (1993). In A. A. Balkema (ed.) *Geochemistry, ground water and pollution* (p. 536). Rotterdam.

Aquachem V. 4.0. (2003). User's Manual. Lukas Calmbach and Waterloo Hydrogeologic, Inc, p. 276.

Berkovwitz, B., Dror, I., Yaron, B. (2008). Contaminant geochemistry, interaction and transport in the subsurface environment (p. 412). New York: Springer.

Committee on Techniques for assessing ground water vulnerability (USA). (1993). *Ground water vulnerability assessment: contamination potential under conditions of uncertainty* (p. 204). Washington: National Academic Press.

COST action 620: Vulnerability and risk mapping for the protection of the carbonate (karst) aquifer. Final report, 2003. (edited by Francois Zwahlen), European Commission, Directorate —General for Research: 297p.

Driver, J. I. (1988). *The geochemistry of natural water* (p. 436). Upper Saddle River: Prentice Hall Inc.

Faure, G. (1998). *Principles and applications of geochemistry* (p. 600). Upper Saddle River: Prentice Hall.

Fetter, C. W. (2001). *Applied hydrogeology* (4th edn., p. 598). Upper Saddle River: Prentice-Hall.

Freeze, R. A., Cherry, J. A. (1979). Groundwater (p. 604). Upper Saddle River: Prentice Hall.

Goldberg, V. M., & Gazda, S. (1984). Gidrogeologicheskie osnovy okhrany podzemnykh vod ot zagryazneniya [Hydrogeological principles of groundwater protection against pollution] (p. 238). Moscow: Nedra.

Goldberg, V. M. (1987). Vzaimosveazi zagreaznenia podzemnyh vod i prirodnoi sredy (p. 248). Moscow: Gidrometeoizdat.

Gurdak, J. J. (2008). Ground-water vulnerability: Nonpoint-source contamination, climate variability, and the High Plains aquifer (p. 223). Saarbrucken, Germany: VDM Verlag Publishing, ISBN: 978-3-639-09427-5.

Heath, R. C. (1987). *Basic ground-water hydrology* (p. 84). USGS: US Government printing office.

Hudak, P. F. (2000). *Principles of hydrogeology* (p. 204). USA: Lewis Publishers.

Langmir, D. (1997). *Aqueous environmental geochemistry* (p. 601). Upper Saddle River: Prentice Hall Inc.

Merkel B. I., Planer–Friedrich B. (2008). *Groundwater geochemistry: A practical guide to modeling of natural and contaminated aquatic systems*, 2nd ed. (p. 230). Springer.

Moraru, C. E. (2009). Gidrogeohimia podzemnyh vod zony activnogo vodoobmena krainego Iugo-Zapada Vostocno – Evropeiskoi platformy (Vol. 1, p. 210). Chisinau: Elena.

Moraru, C. E. & Anderson, J. A. (2005). A Comparative Assessment of the Ground Water Quality of the Republic of Moldova and the Memphis, TN Area of the United States of America (p. 188). Ground Water Institute, Memphis, TN.

Moraru, C., & Zincenco, O. (2005). Podzemnye vody g. Kishinev (Vol. 1, p. 111). Chisinau: Elena.

Myrlean, N. F., Moraru, C. E. & Nastas, G. E. (1992). The Ecological and Geochemical Atlas of the City of Chisinau (p. 191). Chisinau: Stiinta (in Russian).

Perelman, A. I. (1982). Geohimia prirodnyh vod (p. 152). Moscow: Nauka.

Perelman, A. I. (1989). Geohimia (p. 528). Moscow: Vyshaia skola.

Polubarinova-Kocina, P. Ya. (1952). *The theory of ground water movement* (p. 380). Moscow: Nauka.

Samarina, V. S. (1977). *Gidrogeohimia* (p. 280). Leningrad: LGU.

Soliman, M. M., et al. (1997). *Environmental hydrogeology* (p. 386). Boca Raton: Lewis publishers.

Sources of pollution. (2014). http://www.euwfd.com/html/source_of_pollution_-_overview.html.

Shestakov, V. M., & Pozdneakov, S. P. (2003). Geogidrologia. Akademkniga, p. 176.

Tindal, J. A., Kunkel, J. R., & Anderson, D. E. (1999). *Unsaturated zone hydrology for scientists and engineers* (p. 624). United Saddle River: Prentice-Hall Inc.

Todd, D. K. (1980). *Groundwater hydrology* (p. 535). USA: Willey.

Wellings, S. R., & Bell, J. P. (1980). Movement of water and nitartes in the unsaturated zone of upper chalk near Winchester, Hants, England. *Journal of Hydrology, 48,* 119–136.

Wellings, S. R., & Bell, J. P. (1982). Pysical control of water movement in the unsaturated zone. *The Quaternaly Journal of engineering geology, 15,* 235–241.

Wight, W. D., & Sondereger, J. L. (2001). *Manual of applied field hydrogeology* (p. 608). New York: McGraw-Hill.

Chapter 4
Environmental Settings of Study Territories

Constantin Moraru and Robyn Hannigan

Abstract For the estimation of hydrogeochemical vulnerability, representative areas from USA, Germany, and Moldova are characterized environmentally. For the selected areas, general delineation, physiography and climate, soil cover, surface water, geology and hydrogeology, tectonics and paleohydrogeology, and land use are described in details. The natural conditions of groundwater have been provided with a detailed outline.

Keywords Representative areas from USA · Germany and Moldova
Environmental setting · Groundwater

4.1 Memphis Area, USA

4.1.1 General Delineation

The study area consists of Memphis and Shelby County and its surrounding areas to include its local municipalities. Shelby County is geographically located in the State of Tennessee in the USA as illustrated in Fig. 4.1. Geographically, the greater Memphis area is the same as Shelby County.

Shelby County was established in 1819, and Memphis is its county seat. Shelby County honors Isaac Shelby, who was a US Commissioner as was Andrew Jackson who later became the seventh president of the USA from 1829 to 1837. Jackson and Shelby together arranged the purchase of the Western District in 1818 from the Chickasaw Indian Nation. Shelby County has both agricultural and urban land, and there is a local industry. World History (2004) states the following: the county has a total area of 2030 km^2 (784 mi^2) of which 1954 km^2 (755 mi^2) is land and 75 km^2 (29 mi^2) is water. As of the 2015 census, Shelby County's population was 897,472, with 338,366 households and 228,735 families. The population density is 459/km^2 (1189/mi^2). There are 362,954 housing units, producing an average density of 186/km^2 (481/mi^2).

© Springer International Publishing AG 2018
C. Moraru and R. Hannigan, *Analysis of Hydrogeochemical Vulnerability*,
Springer Hydrogeology, https://doi.org/10.1007/978-3-319-70960-4_4

Fig. 4.1 Location of the study area

4.1.2   *Physiography and Climate*

The Memphis and Shelby County area is part of the Gulf of Mexico Coastal Plain. The subdivisions of this regional structure are the Mississippi River Valley and the West Tennessee Plain. The Mississippi River Valley exerts a topographic influence on the area. Altitude ranges from 56.0 m above mean sea level to 131.0 m above mean sea level (The Comprehensive Planning Section 1981). The most striking relief features of Shelby County occur where the upland area terminates, causing 15–46 m high bluffs along the eastern margin of the alluvial plain of the Mississippi River. The resulting bluffs which protected the area from the annual flooding of the Mississippi River were one of the primary reasons for the settlers to form a community in this area. The altitude of about 70% of the territory is between 70.0 and 120.0 m above mean sea level. Geological evolutions formed different types of slopes. There are four categories of slopes:

1. 0–2%—nearly level
2. 2–8%—gentle slopes
3. 8–12%—moderate slopes
4. 12% and above—steep slopes.

The first category of slopes is characteristic of streams; it coincides with floodplains and drainage patterns. Approximately 30% of Shelby County's territory is comprised of nearly level slopes. The second category of slopes is characteristic of older part of the City of Memphis and of surrounding towns like Arlington and Germantown. Gentle slopes also occupy about 30% of the study's territory. The third and fourth categories are geomorphologically connected with elevations more than 90.0 m. These areas are watersheds for the local hydrographic network. Moderate and steep slopes belong to the remaining 40% of the Memphis area.

The climate of the Memphis area is temperate with a mean annual air temperature of about plus 15 °C and average annual humidity of 70%. Winds blow predominantly from the south. Although the area is well inland from large bodies of water, it lies in the path of cold air moving southward from Canada and warm, moist air moving northward from the Gulf of Mexico. Consequently, extreme and frequent changes in the weather are common (Soil Survey, Shelby County 1989). Climatic data are representative for the whole county due to the small variations in altitudes.

Precipitation is the major factor in determining the characteristics of surface and undergroundwater. Area precipitation is abundant and uniformly distributed. Table 4.1 and Fig. 4.2 present statistical data for the Memphis International Airport for the period of 1948–2014 (Precipitation 2014). The annual precipitation value is 1361.94 mm. The maximum atmospheric precipitation is characteristic of the period from November to April, and consequently minimum atmospheric precipitation occurs in the period from May to October.

Table 4.1 Statistical values of atmospheric precipitation, Memphis International Airport

Month	Jan	Feb	Mar	May	Jun	Jul	Aug	Sep	Oct	Nov	Dec	Sum
Precipitation (mm)	116	118	138	129	100	108	84	86	80	126	137	1362

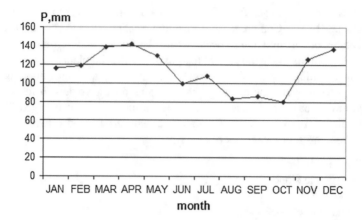

Fig. 4.2 Graphical distribution of atmospheric precipitation (P), Memphis International Airport

Table 4.2 Statistical values for evaporation of precipitations, Shelby County, Tennessee, USA

Month	Jan	Feb	May	Apr	May	Jun	Jul	Aug	Sep	Oct	Nov	Dec
Evaporation (mm)	3	7	22	44	70	90	80	77	60	26	12	5

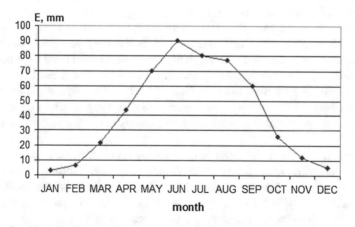

Fig. 4.3 Graphical distribution of evaporation (E), Shelby County, Tennessee

Shelby County's meteorological stations did not record evaporation for a long time period. Table 4.2 and Fig. 4.3 (Evaporation 2014) present extrapolated data from US regional maps.

Analysis of evaporation data shows that its highest values occur from April to September. Analysis of these data with other meteorological data shows that evaporation is controlled by air temperature, relative humidity, and wind velocity. The data presented in Table 4.2 and Fig. 4.3 are statistically reliable because they reflect the regional trend in changing atmospheric parameters (temperature, etc.).

Table 4.3 Geographical data for atmospheric depositional station

Station	Hatchie National Wildlife Refuge (TN14)
Location	Haywood County, Tennessee
Dates of operation	10/2/1984—Present
Latitude	35.4678
Longitude	−89.1586
Elevation	107 m
Distance between Shelby and Haywood counties	87 km

The National Atmospheric Depositional Program investigates the chemical composition of atmospheric precipitation in the USA. The nearest station to the Memphis area is in Haywood County, Tennessee. Table 4.3 (Annual Data 2004) presents data from this station (as the most representative for the period 1984–2017).

The chemical composition of precipitation is changeable in time, but anomalies are not recorded. Table 4.4 shows the average data for the last 20 years.

Variations of water conductivity over time are functionally correlated with total dissolved solids, showing that the chemistry of major ions changes slowly and essentially seasonally as illustrated in Fig. 4.4.

Table 4.4 Chemistry of the atmospheric precipitation, Haywood County, Tennessee, Station TN14

Ingredient	Ca	Mg	K	Na	NH$_4$	NO$_3$	Cl	SO$_4$	pH	Cond.
Value	0.09	0.02	0.02	0.10	0.12	0.86	0.17	1.14	4.69	12.75

Remark chemical elements—mg/l; Conductivity—µS/cm; pH—units

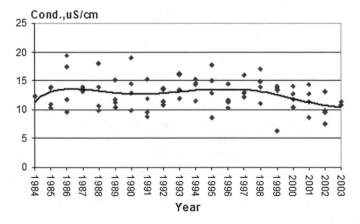

Fig. 4.4 Variation of atmospheric precipitation conductivity, Haywood County, Tennessee, Station TN14 (continuous line is the statistical trend)

Table 4.5 Statistics on selected pesticides in rain and air (Coupe et al. 1999)

Pesticide	Phase	Urban				Agriculture			
		n	Max	Med	%	n	Max	Med	%
Atrazine	Rain	16	0.096	0.006	69	16	0.83	0.02	75
	Gas	24	nd	nd	0	21	42,888	nd	42
	Particulate	24	0.019	nd	29	21	0.42	0.058	67
Chlorpyrifos	Rain	16	0.009	0.005	63	16	0.04	<0.005	38
	Gas	24	42,858	42,856	96	21	42,738	nd	38
	Particulate	24	nd	nd	0	21	Nd	nd	0
Cyanazine	Rain	16	0.074	<0.013	31	16	0.32	0.008	56
	Gas	24	0.61	nd	8	21	0.25	nd	5
	Particulate	24	nd	nd	0	21	0.39	nd	24
Diazinon	Rain	16	0.019	0.005	56	16	0.013	<0.008	13
	Gas	24	42,833	0.14	50	21	42,826	nd	10
	Particulate	24	0.2	nd	25	21	Nd	nd	0
Methyl parathion	Rain	16	0.3	0.024	56	16	43,000	0.12	69
	Gas	24	0.99	nd	46	21	62	42,857	71
	Particulate	24	nd	nd	0	21	0.4	nd	29
Molinate	Rain	16	0.025	<0.004	25	16	0.37	0.026	63
	Gas	24	0.44	nd	4	21	42,828	0.076	62
	Particulate	24	nd	nd	0	21	0.089	Nd	5
Propanil	Rain	16	0.14	<0.016	38	16	42,948	0.036	81
	Gas	24	0.24	nd	13	21	42,893	0.37	57
	Particulate	24	0.043	nd	21	21	42,798	0.54	62
p,p'-DDE	Rain	16	<0.006	<0.006	0	16	<0.006	<0.006	0
	Gas	24	0.19	nd	33	21	42,736	0.67	100
	Particulate	24	nd	nd	0	21	0.01	9 0.01	52
Trifluralin	Rain	16	0.01	<0.002	13	16	0.024	0.007	69
	Gas	24	0.76	0.028	88	21	42,860	0.81	100
	Particulate	24	nd	nd	0	21	0.013	nd	5

Remark [rain units, micrograms per liter; gas and particulate units, nanograms per cubic meter; *n* number of samples; %, percent of sample detections; *max* maximum concentrations; *med* median concentration; *nd* not detected; *n* is the number of samples.]

The state of Tennessee and, in particular, Shelby County is an agricultural area. Pesticides are widely used to protect crops and to increase the comfort and safety of urban residents. Varying amounts of pesticides can be transported long distances through atmospheric air and precipitation. Precipitation can be contaminated with pesticides. Coupe et al. (1999) investigated the occurrence of pesticides in rain and air in Hinds County near the town of Rolling Fork in Sharkey County, Mississippi, about 300 km south of the Memphis area, and results are outlined in Table 4.5 and illustrated in Fig. 4.5.

Similar data for Shelby County do not exist. However, since many of the same crops are found in the area that are similar to those in Hinds County, MS, one can

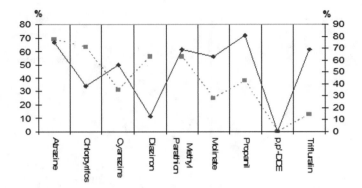

Fig. 4.5 Distribution of pesticides in rain (constructed using data of Coupe et al. 1999) (continuous line—agriculture and discontinuous line—urban areas)

suspect that the presence of the same pesticides in the air and rain may be probable. Table 4.5 and Fig. 4.5 indicate that propanil and atrazine are the most frequently occurring pesticides in rain. These pesticides are widely recognized as toxic substances that are linked to increased cancer rates and chronic illness.

4.1.3 Soil Cover

The seven soil characteristics of Shelby County are detailed below (Soil Survey, Shelby County 1989):

1. Robinsonville–Grevasse–Commerce association nearly level, excessively drained to somewhat poorly drained, loamy, and sandy soils on first bottom along the Mississippi River.
2. Tunica–Sharkey–Bowdre association level, dark-colored, poorly drained to moderately well drained, heavily clay soils on low flood plains of the Mississippi River.
3. Memphis association chiefly steep, well drained, silty soils on uplands.
4. Memphis–Grenada–Loring association nearly level to sloping, well drained and moderately well drained, silty soils on broad uplands.
5. Falaya–Waverly–Collins associations level, poorly drained to moderately well drained, silty soils on first bottoms.
6. Grenada–Galloway–Henry association gently sloping to nearly level or depressional, moderately well drained to poorly drained, silty soils on uplands.
7. Grenada–Memphis–Loring association gently sloping to strongly sloping, moderately well drained to well drained, silty soils on uplands.

The thickness of the soil cover varies from several cm to 2.5 m. Figure 4.6 shows the water balance of the soil horizon (Soil Survey, Shelby County 1989).

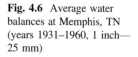

Fig. 4.6 Average water balances at Memphis, TN (years 1931–1960, 1 inch— 25 mm)

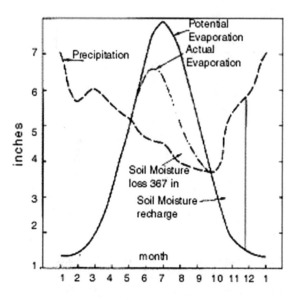

Loess and alluvium were the parent material of the soil in Memphis area, and this determines the geochemical and physical properties of the soil stratum. Permeability of the stratum is 0.005–0.05 m/hr, depth to seasonal groundwater table —0.0–0.6 m and rarely 1.2–3.8 m, and pH values = 5.1–7.8 (Soil Survey, Shelby County 1989). There is no official publication of chemical analyses for soil profiles. Trace elements and pesticides probably occur in soil because of agricultural and transportation practices, as well as the influence of the Memphis International Airport. The upper part of the soil (first 0.5 m) is probably the most polluted. Erosion, the removal of loessial soil, is the main soil limitation in the Memphis area.

4.1.4   Surface Water

Shelby County is bordered on its western edge by the Mississippi River and has three major tributaries trisecting it. Those tributaries are from the south, Nonconnah Creek, Wolf River, and the Loosahatchie River. Horn Lake Creek is a smaller tributary on the southern fringe of Shelby County. Figure 4.7 illustrates the spatial location of the surface waters of the Memphis area of the Mississippi River, its tributaries and local lakes.

The Mississippi River is the second-longest river in the USA and flows from north to south along the western boundary of Shelby County. The Mississippi River, with its large valley, determines the main structure of the surface topography and local hydrologic network of the county. The Wolf River, Loosahatchie River, Nonconnah Creek, and Horn Lake Creek are the four largest tributaries in Shelby County, and each drains westward into the Mississippi River. There are many other smaller streams,

Fig. 4.7 Surface water and topography of Shelby County, Tennessee, USA (Scale 1:200,000 or 1 cm = 2.0 km)

ponds, and lakes scattered throughout the area (The Comprehensive Planning Section 1981). Table 4.6 presents general hydrologic data for the four tributaries.

Figure 4.8 shows that the river network is topographically very dense. This contributes to a fast accumulation of atmospheric precipitation into main rivers and intense drainage of the water table aquifers, mainly from terrace and alluvium deposits. Minimum base flow of the hydrological regime of Shelby County rivers is maintained by shallow groundwater. The usual and high flows are produced by atmospheric precipitation. Table 4.7 shows the average multiannual instantaneous discharge of major tributaries.

Figure 4.9 shows that river discharge over time is very changeable and unstable. Stream flow characteristics essentially correlate with periods of atmospheric precipitations. Water quality in the local hydrologic network is heavily influenced by anthropogenic activity. Figure 4.10 illustrates this fact for the Loosahatchie River.

Data from Fig. 4.10 show that water conductivity seems to be closely correlated with river discharge. Heavy and persistent atmospheric precipitation increases

Table 4.6 Hydrologic data for Mississippi River tributaries, Shelby County, Tennessee (The Comprehensive Planning Section 1981)

Parameter	Loosahatchie river	Wolf river	Nonconnah river	Horn Lake Creek/ Cypress Creek
Total surface area in basin, sq. km	1789.69	1973.58	411.81	119.43
Surface area in Shelby County, sq. km	647.5	533.54	318.57	77.7
Major tributaries in Shelby County	Big Creek Beaver Creek Clear Creek	Cypress Creek Workhouse Bayou Fletcher Creek Grays Creek Mary's Creek	Cane Creek Days Creek Hurricane Creek Ten Mile Creek Johns Creek	Horn Lake Cutoff Horn Lake Creek Cypress Creek

Remark location of the river basins—see Fig. 4.7

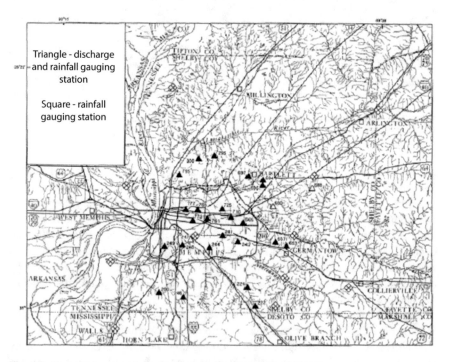

Triangle - discharge and rainfall gauging station

Square - rainfall gauging station

Fig. 4.8 Hydrologic network Shelby County, Tennessee (Neely 1984)

Table 4.7 Statistical parameters of instantaneous discharge (m^3/s) of major tributaries

Parameter	Wolf river	Nonconnah river	Loosahatchie river
Mean	36.44	9.90	9.47
Standard error	6.06	3.97	2.07
Median	13.72	0.08	3.04
Standard deviation	66.90	47.03	23.07
Sample variance	4475.18	2211.99	532.28
Kurtosis	27.21	31.14	21.39
Skewness	4.66	5.57	4.50
Range	526.10	317.15	155.42
Minimum	6.26	0.00	1.92
Maximum	532.36	317.15	157.35
Count data	122.00	140.00	124.00

runoff from river basins. Within a few time lags, increases in water conductivity and river discharge occur simultaneously. Since conductivity is an integral parameter of water quality, water quality changes drastically.

During geological and hydrological evolution, river networks formed a dip erosion structure. The dip basis of erosion is common for the Loosahatchie River, the Wolf River, and the Nonconnah Creek as illustrated in Fig. 4.11. Analysis of available geological and hydrological data indicates that the groundwater of these rivers connect, both partially gaining and losing streams. Losing capacity is probably common for the lower parts of these rivers. This effect is increasing because of the water level drawdown in the Memphis sand aquifer. Hutson and Morris (1992) noted that Nonconnah Creek has a possible hydraulic connection with the Memphis aquifer. In addition, this author has noted other data that confirm seepage from Nonconnah Creek into the Memphis aquifer.

4.1.5   Geology and Hydrogeology

4.1.5.1   General Geology

Many previous reports and publications include information concerning the geology of the Memphis area. Kazmann (1944), Schneider and Cushing (1948), Cushing et al. (1964), Boswell et al. (1965), Hardeman (1966), Parks (1973), Parks et al. (1981), Saucier (1994), and Hosman (1996) describe important data.

The Memphis area is a small part of the regional geological structure called the Mississippi embayment syncline, and it is located in the northwest part of the embayment. The geological history of the embayment has strongly influenced all of the geological particulars of the study area. Two floors of rock comprise the geological structure of the area. The first is a bedrock floor of the Paleozoic age and consists of limestone. The top of the Paleozoic rocks is between 670 and 980 m

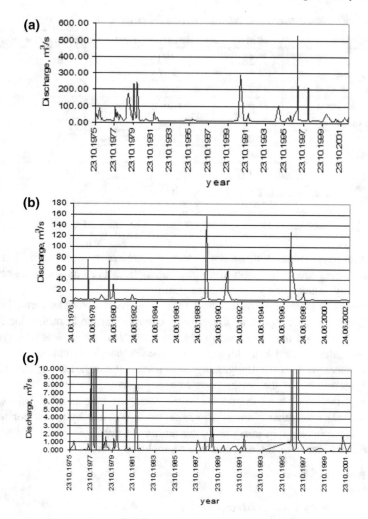

Fig. 4.9 Representative average monthly discharges of major tributaries (**a** Wolf River, **b** Loosahatchie River, **c** Nonconnah Creek; for Nonconnah Creek maximum values are greater than 10 m³/s)

below sea level. The slope of the bedrock, about 5 m per 1 km, is toward the axis of the embayment and inclination, in the region of Shelby County. The second floor ranges in time of origin from the Cretaceous through Recent Quaternary. This is a complex sandwich of clay, silt, sand, and gravel, with a total thickness of about 915 m. Table 4.8 presents a generalized stratigraphy of the Memphis area. Figure 4.12 is a geological map, and Fig. 4.13 is a generalized cross section.

Deposits of the Cretaceous system in the perimeter of the Memphis area are present only by an upper series of Upper Cretaceous. These deposits rest uncomfortably on the Paleozoic rocks. Sediment is of marine origin and ranges from clay

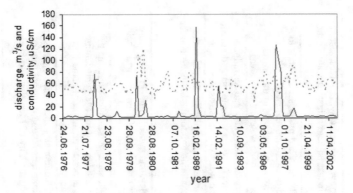

Fig. 4.10 Average monthly discharge and water conductivity (Loosahatchie River) (continuous line is discharge, and discontinuous line is conductivity)

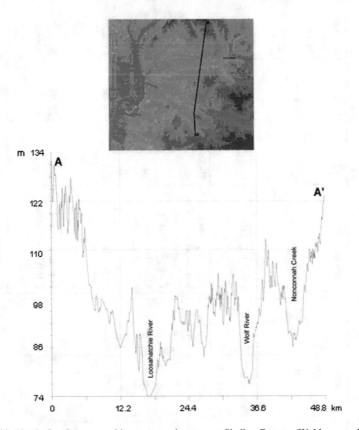

Fig. 4.11 North–South topographic cross section across Shelby County (Waldron et al. 2011) (A–A' is cross-sectional line)

Table 4.8 Post—Paleozoic geologic units underlying the Memphis area (after Brahana and Broshears 2001 and modified by authors)

System	Series	Group	Stratigraphic unit	Geological index	Thickness (m)	Hydrologic unit	Lithology and hydrogeology
Quaternary	Holocene and Pleistocene		Alluvium	Qal	0–53	Shallow (phreatic) aquifer	Sand, gravel, silt and clay. Provides water to farms, industrial, and irrigation wells in the Mississippi Alluvial Plain
	Pleistocene		Loess	Ql	0–20		Silt, silty clay, and minor sand. Tends to retard downward movement of water-providing recharge to the fluvial deposits. Provide water in rural areas
Quaternary and tertiary (?)	Pleistocene and pliocene (?)		Fluvial deposits (terrace deposits)	Qt	0–30		Sand, gravel, minor clay and ferruginous sandstone. Provides water to many domestic and farm wells in rural areas
Tertiary	Eocene	? —	Jackson formation and upper part of the Claiborne group ("capping clay")	Tj–Tc	0–113	Confining unit	Clay, silt, and minor lignite. Serves as the upper confining unit for the Memphis aquifer
	? —	Claiborne	Memphis Sand ("500-foot" sand)	Tc	152–271	Memphis aquifer	Sand, clay, and minor lignite. Primary source of water for the city of Memphis and Shelby County
	Paleocene	Wilcox	Flour Island formation	Tw$_1$	43–94	Confining unit	Clay, silt, sand, and lignite. Serves as the lower confining unit for Memphis aquifer, and the upper confining init for the Fort Pillow aquifer
			Fort pillow sand ("1400-foot" sand)	Tw$_2$	28–93	Fort pillow aquifer	Sand with minor clay and lignite. Sand is fine to medium. The second principal aquifer supplying the City of Memphis and Shelby County
			Old breastworks formation	Tw$_3$	55–107		Clay, silt, sand, and lignite. Serves as the lower confining unit for the Fort Pillow

(continued)

Table 4.8 (continued)

System	Series	Group	Stratigraphic unit	Geological index	Thickness (m)	Hydrologic unit	Lithology and hydrogeology
		Midway				Midway confining unit	aquifer, along with the underlying Porters Creek Clay, Clayton formation, and Owl Creek formation
			Porters Creek Clay	Tm_1	76–97	Midway confining unit	Clay and minor sand. Thick body of clay with lenses of clayey, glauconitic sand. Principal confining unit separating the Fort Pillow aquifer and the Ripley Formation and McNairy sand
Cretaceous	Upper cretaceous		Owl Creek formation	Ko	12–27	Midway confining unit	Clay and sand. Calcareous clay and glauconitic sand; fossiliferous. Confining unit
			Ripley formation and McNairy sand	Km	110–174	McNairy–Nacatoch aquifer	Sand and clay, minor sandstone, limestone, and lignite. Aquifer with low potential for use in Memphis area because of lesser amount of sand and poorer quality of water than aquifer above
			Coon Creek formation	Kcc	0–18	Confining unit	Clay and sand. Probably present only in northeastern Shelby County. Confining unit
			Demopolis formation	Kd	82–119	Confining unit	Clay and chalk, glauconitic and fossiliferous. Serves as the principal confining unit separating the ripley formation and NcNairy sand and coffee sand
			Coffee sand	Kc	0–37	Coffee aquifer	Sand and minor clay. Sand fine to medium, locally glauconitic or lignite. Clay occurs mainly at the basis. Contains brackish or saline water; not considered a freshwater aquifer in the Memphis area

(continued)

Table 4.8 (continued)

System	Series	Group	Stratigraphic unit	Geological index	Thickness (m)	Hydrologic unit	Lithology and hydrogeology
Ordovician		Richmond and Nashville (?)	Nannie Shel, Fernvale Limestone and Hermitage formation (?)	O	More than 25	Upper part as confining unit	Mainly limestone. Confining unit in the upper part separating the coffee aquifer; deeper as aquifer containing saline water

Fig. 4.12 Geological map of Shelby County, Tennessee (scale 1:250,000 m; light gray—Quaternary and dark gray—tertiary) (Hardeman 1966)

to different sands (see Table 4.8). The top of the Cretaceous system in the region of this study area is between 365 and 610 m with an inclination following the Paleozoic surface, i.e., from northeast to southwest.

The Tertiary system combines deposits of the Paleocene, Eocene, Pliocene, and Pleistocene series. The Midway rock group is the lower part of the Paleocene, and it lies uncomfortably over the Cretaceous system. Clay and sand of marine origin are the predominant sediments (see Table 4.8). After the Midway rock group, the Wilcox group of Paleocene–Eocene series follows in the cross section. There is a heavy parting of deposits, consisting mainly of sandwiched clay, sand, silt, and lignite. The sediments are mainly of marine origin but in very shallow conditions, a hot climate, and organically rich environment. The Claiborne group of sediments belongs to the Eocene series. These rocks outcrop in the eastern part of Shelby County. Lithologically, these rocks consist of a variety of sands, clays, silts, and minor lignites (see Table 4.8). These rocks are of marine and nonmarine origin (Cushing et al. 1964). Pleistocene and Pliocene rocks are undifferentiated within West Tennessee including the Memphis area. Fluvial deposits are the main rocks of this age. The Quaternary system covers all of Shelby County as illustrated in Fig. 4.12. These sediments include sand, gravel, silt, and clay. The most significant sediment is ferruginous sandstone, whose origin is not well understood.

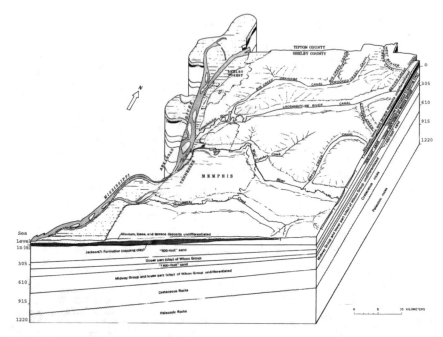

Fig. 4.13 Block—diagram showing physiographic and geological features, Memphis area, Tennessee (Edwin and Nyman 1968)

4.1.6 Hydrogeological Conditions

The City of Memphis ranks second after San Antonio, Texas, among the American cities that depend solely on groundwater for its municipal water supply (Brahana and Broshears 2001). Such a situation is rare in international practice. For this reason, there are a substantial number of publications on the hydrogeology of the Memphis area. The most important publications for the present study include those of Flohr et al. (2003), TDEC (2002), Gonthier (2002), Brahana and Broshears (2001), Kleiss et al. (2000), Robinson et al. (1997), Parks (1990), Arthur and Taylor (1990), Kingsbury and Parks (1993), Hutson and Morris (1992), McMaster et al. (1988), Brahana et al. (1985), (1986), (1987), Parks et al. (1981), Edwin and Nyman (1968), Hosman et al. (1968), Boswell et al. (1965), (1968), Nyman (1965), Criner and Armstrong (1958), and Kazmann (1944).

The first aquifer from the land surface is the shallow aquifer. Water-bearing rocks are alluvium, loess, and fluvial deposits as presented in Table 4.8. These three stratigraphic units have one common groundwater level and near identical water geochemistry water. Therefore, one unique aquifer joins Quaternary deposits which is characteristic for flood plain in Pleistocene–Holocene alluvium, for terraces in Pleistocene, and Pliocene (?) fluvial deposits, and for Pleistocene loess on highlands. The shallow aquifer is unconfined but is locally confined by the loess silt in the western part of the study area where the water level is above the top of the sand

and gravel (Gonthier 2002). The Jackson formation and the upper part of the
Claiborne group are considered the confining unit for the phreatic aquifer. Brahana
and Broshears (2001) mapped the thickness of the Jackson—Upper Claiborne
confining unit in the Memphis area. The thickness varies from 0 to 90 m, but is
predominantly 30–60 m. In some places, the thickness of the confining unit is less
than 3.0 m, or practically absent. Such spaces act as "lithological windows." It
permits the existence of a hydrodynamic relationship between the shallow and
Memphis aquifer as well as a potential probability of water leakage from up to
down in the geological cross section. At present, "windows" are not precisely
mapped and different opinions exist concerning their location and methods of
detection (Larsen et al. 2002). Figure 4.14 shows the water table contours of the
shallow aquifer. From this map, it is evident that movement of phreatic water
follows the land surface topography. Recharge areas are located on highlands, and
discharge takes place in the river network. The main basis of discharge is the
Mississippi River valley, and the main phreatic water streams are oriented in this
direction. The water table is very changeable over time. As a rule, a direct correlation between water level and atmospheric precipitation is difficult to find. This
property is a universal characteristic in many countries. We propose to correlate the
differences (atmospheric precipitation–evaporation) and water levels. This correlation is illustrated in Fig. 4.15. Data used for this graph area (a) area random time
interval, and one correlation is common for all of the monitoring periods. As
illustrated in Fig. 4.15b, a, simple correlation between water level and precipitation
does not appear to have a clear statistical aspect. After transformation of

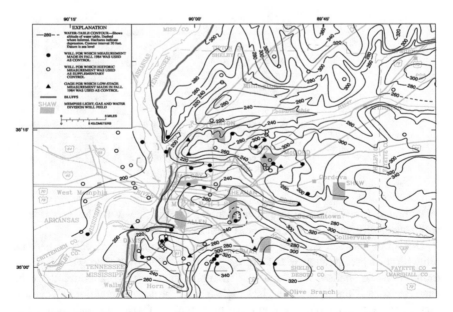

Fig. 4.14 Water table contours of the shallow aquifer (Brahana and Broshears 2001; unit is feet;
1 feet = 0.3048 m)

Fig. 4.15 Relationship between atmospheric precipitations (P), atmospheric precipitation—evaporation (P − E) and water level in selected well Sh: P − 99 (**a** well location, **b** relation P-Water level and **c** relation P − E, and water level; continuous line—precipitation, and difference precipitation—evaporation; discontinuous line–water level in the monitoring well)

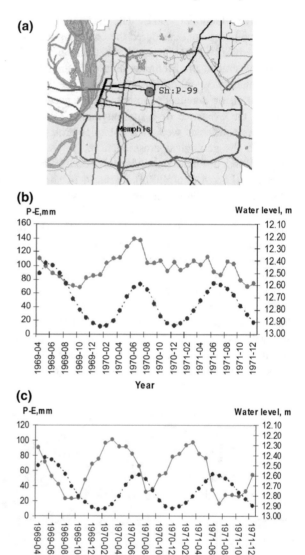

precipitation (precipitation minus evaporation, P − E), the water level correlates well statistically with the unit P − E as seen in Fig. 4.15c.

Deviation between extreme values of water level and P − E occurs because infiltration of precipitation is not prompt but distributed over time. For all monitoring wells of shallow aquifer in the Memphis area, analogous correlations are characteristics if the hydrogeological regime is in the natural condition. This distinguishes the fact that precipitation is the main source of water for shallow aquifers and its infiltration through unsaturated zone is rapid.

The water quality of the shallow aquifer is a function of evaporation of the water table, the concentration of salts, and anthropogenic impact on the water table. In addition, as a general practice, water from this aquifer is not used in the Memphis metropolitan areas as drinking water. Hydrodynamic parameters such as the hydraulic conductivity and the transmissivity are generally unknown. In rural areas, this water is used for small domestic purposes and irrigation.

The second aquifer is in the water-bearing rocks of the Claiborne group as shown in Table 4.8 and Fig. 4.13. The local name of this aquifer is "500-foot" sand or Memphis sand, the latter name being used by the US Geological Survey. Memphis sand is a regional aquifer in the region of the Mississippi embayment. It is known as the Sparta sand (Arthur and Taylor 1990) in southern Arkansas, Louisiana, and Mississippi. In the Memphis area, Criner and Armstrong (1958) divided this aquifer into lower and upper parts. The lower part contains more clay beds that are thicker and more extensive than those in the upper part. The thickness of the clay beds is not uniform, so no particular bed can be considered to be a hydrologic boundary. The "500 foot" sand is considered to be a single hydrologic unit (Criner et al. 1964). Geophysical logs have been used to delineate the top and bottom of the aquifer. In most cases, it is not possible to show the distinct altitude of the top parts of the Jackson formation that are included as water-bearing deposits. Nevertheless, values of the aquifer top with respect to mean sea level are between +92 m (Collierville, well M-1) and −38 m (Memphis, well O-54); values of the aquifer bottom are about −61 m (Collierville) and −244 m (Memphis) (Criner et al. 1964). The total thickness of the Memphis aquifer is 152–271 m (Table 4.8). The aquifer is inclined to the Mississippi River, which coincides with the axis of the embayment. In most of the study area, the aquifer is confined, but it begins to become unconfined in the southeastern and eastern part of Shelby County. There has been insufficient study of hydraulic properties, and only generalized data are available as follows: Transmissivity is 250–5016 m²/day, and the storage coeffi-cient is 1×10^{-4} to 2×10^{-1} (Brahana and Broshears 2001). The Memphis aquifer is the main source of freshwater for the City of Memphis and other municipalities in Shelby County. Use of water pumping for drinking water began in 1898 and has increased each year since as illustrated in Fig. 4.16. The pumpage rates in Fig. 4.16

Fig. 4.16 Memphis municipal pumpage, 1986–2010 (modified from Arthur and Taylor 1990)

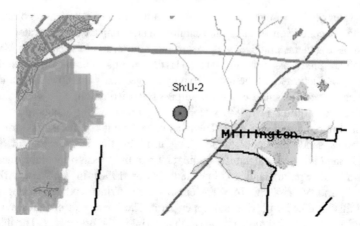

Fig. 4.17 Well location, Memphis aquifer

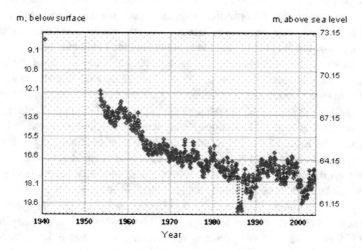

Fig. 4.18 Hydrograph of water level in the Memphis aquifer (well Sh: U-2)

are those that were officially registered. In addition to these volumes of water, many individual wells exist, for which data are not available.

The high rate of water extraction from the Memphis aquifer forms an impressive and regional cone of water depression. Figures 4.17 and 4.18 show an example of drawdown of the water level (USGS 2003).

Hydraulically, the Memphis aquifer is connected on the top with shallow water and a river network that includes the Mississippi River, and the trough on the bottom is connected with the Fort Pillow horizon. The main recharge is received in the eastern part of the county and beyond where the Memphis sand is in outcrop. Water of this aquifer is used by the City of Memphis and the local

Table 4.9 Calculated water budget for the Memphis aquifer, Memphis area (modified by the authors)

Sources and discharges	Flow (m³/sec)	% of total
Sources		
Recharge	3.00	36
Boundary flux	0.48	6
Leakage from shallow aquifer	4.44	54
Leakage from deep aquifer	0.06	1
Storage	0.28	3
Total	8.26	100
Discharge		
Boundary flux out	0.08	1
Pumping	8.18	99
Leakage (net in)	0.00	0
Total	8.26	100

municipalities. Table 4.9 presents the water budget calculated by Brahana and Broshears (2001).

Analysis of the data from Table 4.9 shows that pumping is nearly equal to the sources of water of the horizon. Also, the recharge is drastically increased compared with steady-state conditions. In order to prevent over-exploitation, water extraction from the Memphis aquifer should be reduced and more closely monitored.

The third aquifer in the Memphis area is the Fort Pillow aquifer. The local name is "1400-foot" sand. Interpretation of electric and gamma-ray logs delineates the horizon, and sometimes the upper and lower boundaries are approximations. The Flour Island Formation confining unit (thickness 43–93 m) is located on the top of the aquifer, which basically consists of clay with minor silt, sand, and lignite (see Table 4.8). The Midway confining unit (thickness 55–107 m) is located in the base of the aquifer. Thus, the confining formation is not strongly impermeable and in geological terms can be considered as semi-permeable. Criner et al. (1964) noted that the aquifer is continuous throughout the area and its dip is toward the embayment axis at a rate of about 5 m per km. The thickness of the aquifer is from 28 m in the eastern part to 93 m in the western part. The Fort Pillow sand outcrops east of Shelby County where the recharge zone is located.

In the region of the Memphis area, the Fort Pillow aquifer supplements the water supply of some local, social, and industrial users. Water extraction from this aquifer began around 1925 and averages at present about 0.038×10^6 m³/day. This is approximately 1/17 of the Memphis aquifer. The decrease in the water level of the Fort Pillow aquifer is not so significant (USGS 2003) (Fig. 4.19).

Only very general studies of the hydrodynamic parameters of the water-bearing rocks (mainly fine to medium sand) have occured. Brahana and Broshears (2001) reported the following values: transmissivities 250–1765 m²/day, hydraulic conductivity 9 m/day (only one value), and storage coefficient 0.002–0.02. This variability of parameters of aquifer tests indicates that the Fort Pillow aquifer is not

(a)

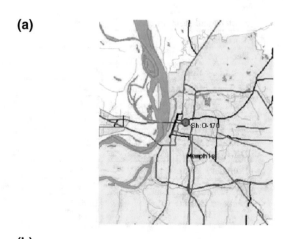

(b)

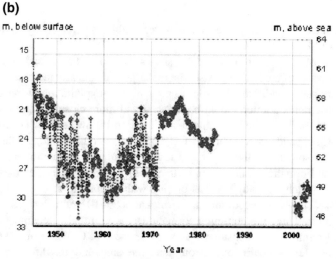

Fig. 4.19 Location (**a**) and hydrograph of water level in the Fort Pillow aquifer (**b**)

homogeneous and contains randomly distributed clay beds. Table 4.10 presents the water budget calculated by Brahana and Broshears (2001).

Data from Table 4.10 indicate that water reserves of the Fort Pillow aquifer are much less than those of the Memphis aquifer (0.45 and 8.26 m³/sec, respectively). Furthermore, nonintensive pumping from the Fort Pillow aquifer is equal to 88% of the total of natural sources.

The McNairy–Nacatoch aquifer is the successive hydrologic unit in the cross section of the Memphis area. This aquifer encompasses deposits of the Ripley formation and McNairy sand, which are overlain by the Midway confining unit and Own Creek formation (total thickness is 143–231 m) and are underlain by the complex confining unit of the Coon Creek and Demopolis Formations. Data about this aquifer for Shelby County are poor. The following data are extrapolated from

Table 4.10 Calculated water budget for the Fort Pillow Aquifer, Memphis area (modified by the authors)

Sources and discharges	Flow (m³/sec)	% of total
Sources		
Recharge	0.14	31
Boundary flux	0.25	56
Leakage from shallow aquifer	0.00	0
Leakage from deep aquifer	0.00	13
Storage	0.06	13
Total	0.45	100
Discharge		
Boundary flux out	0	0
Pumping	0.40	88
Leakage to the Memphis aquifer	0.06	12
Total	0.45 (rounding)	100

the test well, Lauderdale County, Tennessee (Moore and Brown 1969). In the interval 716–850 m is a deposit of gray sand with a lens of clay. Boswell et al. (1965) noted that the Ripley formation thickens toward the axis of the embayment and reaches a maximum thickness of about 175 m in Shelby County. On the territory of the Memphis area, one well in the Mallory well field has been drilled to the McNairy–Nacatoch deposits. Brahana and Broshears (2001) reported that the static water level in this well is about 106 m above sea level and approximately 30 m from surface. Seasonal variation in the water level is 0.6 m, and no long-term decline is evident. The head values of the aquifer are approximately 55 m higher than the heads in the Fort Pillow aquifer. The McNairy–Nacatoch aquifer has not been used to supply the Memphis area with water.

The Coffee aquifer is the last hydrologic unit of the post-Paleozoic geological deposits in the Memphis area. Direct data are not available. By extrapolating data from a test well in Lauderdale County, Tennessee (Moore and Brown 1969), it can be demonstrated that the top of the aquifer is 945 m, and the bottom is 960 m from the ground surface. Sand presents water-bearing rocks that are light gray to light olive-gray, with calcareous cement,glauconitic, pyritic, clear to milky to iron stained quartz grains, loose to friable. The occurrence of this aquifer on the territory of the Memphis area is questionable. The Coffee aquifer is not detected in the deep well Sh: U-12, which is located in the study area. The geophysical log of this well indicated the occurrence of igneous rock in the interval of the Coffee sand (?) (Moore and Brown 1969).

4.1.7 Tectonics and Paleohydrogeology

Geologically, the Memphis area is a small part of the Mississippi embayment. The tectonics and paleohydrogeology of the study territory have the same characteristics

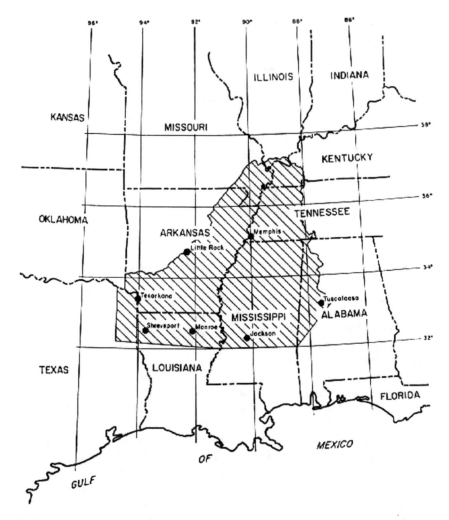

Fig. 4.20 Geographical location of the Mississippi embayment (dash area is the embayment; scale 1:10 km)

that are typical of the entire embayment, especially the northern part. The Mississippi embayment measures about 259,000 km^2 with its location shown in Fig. 4.20 (Cushing et al. 1964).

The current Mississippi embayment is a syncline plunging to the south (Cushing et al. 1964) and forms a northward excursion of the Gulf of Coastal Plain province (Cox and Van Arsdale 2002). The axis of the syncline roughly follows the present course of the Mississippi River. The embayment dates from late Cretaceous to early Tertiary. It is filled with sedimentary rocks ranging in age from Jurassic to Recent

(Cushing et al. 1964). The central and northern parts contain sediments dating only from Late Cretaceous to Recent.

The origin of the embayment is open to question, but Cushing et al. (1964) have proposed the following: The initial tectonic movement that resulted in the formation of the Mississippi embayment syncline may have been a part of the Appalachian evolution at the end of the Paleozoic Era. Other subcrustal movement accompanied the uplift of the Appalachian Mountains, and the initial subsidence in the south-western part of the embayment may have occurred at this time. At the time of the Late Cretaceous and under the influence of the Ouachita Mountain system, the development of the embayment terminated. Saucier (1994) supports this idea, and his interpretation is that continental rifting first created the Gulf of Mexico Basin in the Late Triassic or early Jurassic times. Afterward, crustal downwarping during the Cretaceous epoch along the north side of the Gulf Basin formed the Mississippi embayment, a northward synclinal projection of the Coastal Plains that lay between the Southern Appalachian and Ouachita Mountains. Cox and Van Arsdale (2002) argued that the Mississippi embayment is the result of the Cretaceous superplume mantle event. In the early Cretaceous epoch, the Bermuda hotspot strengthened at the beginning of the superplume event, approximately 120 Ma. In the mid-Cretaceous epoch, the hotspot was beneath the Mississippi Valley Graben. The weak lithosphere lifted and contemporaneous erosion reduced this highland. In the early Cenozoic epoch, this territory moved away from the Bermuda hotspot, and by the late Eocene epoch, subsidence of the embayment was complete.

Van Arsdale and TenBrick (2000) studied the northern part of the Mississippi embayment and concluded that: (1) unlike the commonly portrayed vertically exaggerated bowl-shaped trough, the Paleozoic nonconformity and overlying Upper Cretaceous and Cenozoic sediments are nearly flat; (2) there is no Late Cretaceous vertical faulting; (3) extensive faulting occurred during the Paleocene and lower Eocene eras; (4) extensive and compressive faulting occurred from the middle to late Eocene era; and (5) compressive faulting has and continues to occur during the Quaternary era. Nevertheless, tectonic faults are not very expressive in the soft sediments, and their detection and mapping are still not settled. Kingsbury and Parks (1993) studied the relation of Memphis area faults to interaquifer leakage. They identified faults (or fault zones) that generally have northeast–southwest and northwest–southeast strikes. Vertical displacement along these faults generally ranges from 15 to 45 m for both Memphis sand and the Fort Pillow aquifer. Tectonic faults divide the Memphis area into several blocks. The same authors noted that the relative uplift of blocks bounded by faults might have resulted in erosion of the confining unit, thereby creating windows where the water table aquifers may directly overlie the Memphis aquifer where the potential for down-ward leakage of water from the shallow aquifer into the Memphis aquifer exists. Faults may also control the shape and size of these windows.

The geological history of the Mississippi embayment is complex and continu-ous. During the Cretaceous epoch, the sea reached its maximum northern limit of the Mesozoic Era (Cushing et al. 1964). During the Late Cretaceous epoch, sea transgression from southeast to northwest formed the Coffee Formation which transgression overlies the Paleozoic nonconformity. A high and stable sea level coupled with a warm climate created conditions necessary for sedimentation of the Demopolis and Coon Creek Formations. These formations are mainly clay and minor sand and serve as confining unit for the Coffee aquifer. Primary water in the aquifer originated during sedimentation and water salinity was near marine condition. Perhaps, the water type was sodium chloride with high concentrations of iodine, boron, and other trace elements—characteristic of a marine environment. Gradually, the sea level decreased and its fluctuation together with the intense flushing of the desegregated surface rock material created the Ripley and McNairy formations. These formations consist mainly of sand, local sandstone, limestone, and a lens of clay. For a short time, the sea level rose, and its stabilization resulted in the sedimentation of the Owl Creek formation comprised of clay with secondary sand beds, and many remains of marine biological life. This stratum of clay caps the underlying sand formation. At the same time, the McNairy–Nacatoch aquifer was formed. Without question, the primary origin of water in the aquifer was marine with high salinity and predominantly sodium chloride type.

Cushing et al. (1964) believes that a series of cyclic inundations (transgressions) and regressions of the sea characterize the Tertiary Period. Uplift of the territory and sea regression forms the stratigraphical nonconformity between the Late Cretaceous epoch and the Paleocene epoch. Soon, the sea again covered the northern part of the embayment. In marine conditions, the Midway group of sediments formed, which thins to about 160 m (Van Arsdale and TenBrick 2000) or 76–97 m (Brahana and Broshears 2001). This formation continues to build the confining unit of the previous aquifer. It consists of marine steel-gray to dark-gray clay with disseminated organic material and pyrite inclusions; it is very glauconitic near the base (Van Arsdale and TenBrick 2000). The placidness of the sea and dry climatic conditions permitted the deposit of the Old Breastworks clay formation, a 55–107 m package of impermeable rocks. Gradually, the territory suffered from uplifting and consequently the sea began to regress. The northern part of the embayment and isolated lagoons remained shallow. At the same time, abundant rains formed runoff flows on the land's surface. This transported large quantities of detritus rock material that deposited the Fort Pillow sand. Fort Pillow sand is 28–93 m (Brahana and Broshears 2001) or 64 m thick marine sand at Shelby County that thins and grades to a fluvial/deltaic sand northward, where it is 32 m thick beneath New Madrid (Van Arsdale and TenBrick 2000). Changes in the humid climate to dry conditions helped deposit the Flour Island Formation of impermeable sediments, which are of clay mixture, silt, and sandy clay. The decomposition of flora leads to lignite formation as thin beds. As a practical matter, the Flour Island Formation isolated the Fort Pillow sand and transformed it into the aquifer. The origin of water in the aquifer is marine and fluvial–deltaic. The salinity of the water was lower when compared to seawater, and perhaps, the water type was calcium/magnesium chloride.

The Claiborne group of rocks formed in marine lagoon and fluvial–deltaic conditions. Memphis sand is more fluvial–deltaic sand that is 152–271 m thick in the region of Shelby County. The horizon contains fine to very coarse-grained gray-white sand, lignite, and pyrite inclusions, as well as a high content of organic material. The overlying Jackson formation is also a mixture of marine, fluvial–deltaic clay, silt, and minor lignite with a maximum thickness of 113 m. This formation is a "capping clay" for the Memphis aquifer. The water is primarily moderate saline with high concentrations of iron and manganese. The water type was in a range between calcium/magnesium chloride and sodium sulfate.

By the end of the Tertiary Period, most of the embayment was filled with sediments. Quaternary deposits of sediments occurred in the marine, fluvial, deltaic, lacustrine, and eolian environment. These processes together with surface weathering and erosion formed sequences of water-bearing and impermeable horizons. Accumulation of atmospheric precipitation resulted in the formation of the shallow aquifer. The water quality constantly changed under the influence of climatic conditions and glacial periods. The most significant geochemical fact was the formation of ferruginous inclusions in the coarse sand and gravel of the Lafayette formation of the Pleistocene era. The water type is nearly the same as present conditions. Probably in early Quaternary times, i.e., the beginning of the Pleistocene era, water was brackish and contained high concentrations of iron, manganese, and boron.

Structurally, the Mississippi embayment is a hydrogeologic basin consisting of six regional aquifers. The minimal thickness of all horizons is characteristic of the northern part of the embayment and maximal thickness for the zone south of the syncline. In addition, the thickness of the horizons decreases in the direction of flanges of the embayment, and in the northwest part, the Cretaceous and Tertiary formation outcrops. Overall, such a system has a good hydrodynamic gradient for water movement from flanges to the axis of syncline. Generally, the direction of underground flow is from north to south, and it discharges regionally to the Gulf Coastal Plain. Historically, such natural properties of the basin help to wash down in several cycles all water-bearing permeable formations. Physically, the intensity of washing decreases from Eocene to Cretaceous deposits. The main source of water for this process was melting of continental glaciers. Saucier (1994) described the glacial chronology and glacial cycles of the Mississippi Valley in detail. Eight major glacial cycles are recognized that occurred during the Quaternary Period. Glacial melt forms the hydrologic system of the Mississippi River. In addition, water slowly infiltrates water-bearing horizons. As a result, initial groundwater was substituted and sediments were washed down of marine salts. Additionally, a voluminous amount of precipitation joined with the glacial melt.

New data about groundwater flow confirm the possibility of cyclic washing of aquifers in the Mississippi embayment (Arthur and Taylor 1990). Simulations indicate that the greatest amount of aquifer recharge under predevelopment conditions was to the Memphis aquifer in northern Mississippi and southern Tennessee, where recharge rates exceeded 25 mm/year. Large aquifers transmissivity, high heads in outcrop areas, and short flow paths from recharge to discharge areas contributed to the high rates of recharge and discharge in the northern area of the embayment. The same authors noted that total predevelopment horizontal flow southward across the 35th parallel from the northern area was about 25,326 m^3/day;

the total predevelopment horizontal flow westward across the axis of the embay-ment south of the 35th parallel was about 74,088 m^3/day. Such a significant amount of water reflects, first, the permeability of aquifers and, second, the reality of cyclic washing of the water-bearing strata. Without question, the recent water quality of the Cretaceous and Tertiary aquifers, which completely changed from its condition, demonstrates these considerations.

4.1.8 Land Use

Shelby County is a modern and developed area. Land use is very intensive and mainly in the following categories (in the order of priority): residential, row crops/small grains, commercial/industrial, forest, and others as illustrated in Fig. 4.21. High transportation and pesticide application are two characteristic aspects of land use practices in the area. Transportation includes both ground and air facilities. Auto vehicles and railroads are the important modes of ground transportation. The density of auto vehicles per unit is high, i.e., 186 cars/km^2 (private cars only). Five railroads lines converge on the City of Memphis, and the density of trains is approximately 1 train/hour. A plane lands or takes off every two minutes at Memphis International Airport or the Federal Express airport. Overall, transporta-tion has a huge ecological impact on the habitat, through soil, and groundwater of the Memphis area, especially in light of the city's relatively small territory. However, no direct data exist concerning this assumption. International experience shows that in such an area, a large spectrum of air and soil pollutants is common. Figure 4.22 presents investigations conducted in Russia and Moldova that demonstrate that cities from these two territories are highly polluted by a large variety of toxic trace elements (Myrlean et al. 1992). Beyond question, trace ele-ments are found in human blood, urine, hair, and nails, as well as in vegetables, and in the leaves of trees. It is the author's expert opinion that there should exist a high probability of occurrence of the same pollutants in the Memphis area.

Pesticide application in the study area has occurred for many years. Different pesticides are used in both agricultural and urban areas. Statistically, the quantities of pesticides applied in agriculture every year are the same as in the 1990s. In 1995, Gonthier (2002) reported the following data for Shelby County: Almost 58,000 kg of herbicides was used; the most frequently used pesticide was fluometuron, 7300 kg of which was used for cotton; the second most heavily used pesticide was metolachlor, 6700 kg of which was used for soybeans, cotton, corn, and sorghum; the third most heavily used pesticide was trifluralin, 6500 kg of which was used for cotton and soybeans; more than 11,800 kg of insecticides was used, including 3600 kg of aldicarb; about 3400 kg of fungicides was used. Gonthier also noted that in urban areas, private homeowners use pesticides around their houses and in their gardens. The most heavily used pesticides are acifluorfen, 2,4-D, glyphosate, MCPP, and atrazine. Additionally, Gonthier noted that pesticides have been detected in the local soil cover and in the shallow aquifer.

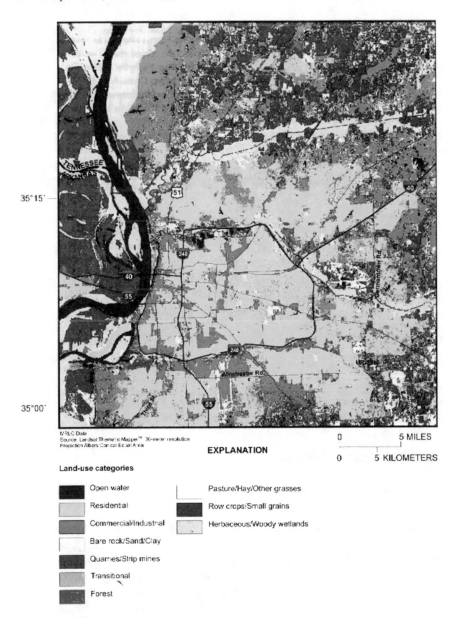

Fig. 4.21 Land use, Shelby County, Tennessee (Gonthier 2002)

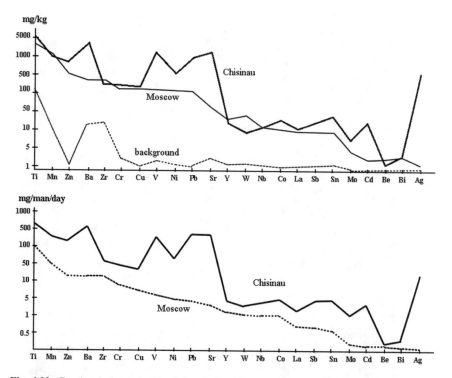

Fig. 4.22 Geochemical spectrums of the air dust due to anthropogenic impact, cities Moscow (Russia) and Chisinau (Moldova) (above graph is the general concentration of trace elements; below graph is daily impact of trace elements to one city man)

4.2 Republic of Moldova

4.2.1 General Delineation

The Republic of Moldova (old name Basarabia) dates from the twelfth century. Its history intertwines with Romania and the former Soviet Union. Until 1918, Moldova was part of the southeastern part of Russia. Between 1918 and 1940, Moldova was part of Romania. From 1940 to 1991, the country was part of the Union of the Soviet Socialist Republics, i.e., the Moldavian Soviet Socialist Republic. Moldova became independent after the collapse of the former Soviet Union in 1991. Moldova is now a member of the Newly Independent States (NIS).

The Republic of Moldova is situated in southeast Europe, east of Romania and north of the Black Sea as illustrated in Fig. 4.23. Moldova is a small and nearly landlocked country with an area of 33,800 km² and a population of about 5 million. Chisinau is its capital and largest city. Moldova shares borders with Ukraine and Romania. The Prut River forms the boundary between Moldova and Romania. Its

Fig. 4.23 Geographical location of the Republic of Moldova

basin on the Moldovan side of the river occupies an area of roughly 8300 km^2, with a population of approximately 1.2 million people.

Moldova has one of the highest population densities in Europe with 129.4 people per km^2. About 47% of the population lives in urban areas, and 53% lives in rural area. Moldovans (Romanians) form a majority in the country (64.5%), followed by Ukrainians (13.8%) and Russians (13%). The Gagauzes in the south are of Turkish origin and constitute 3.5% of the population, and Bulgarians constitute about 2%. Romanian is spoken by 62.7% of the population, Russian by 21.8%, and Ukrainian by 9.8%. The population resides in 21 cities and towns, 48 city-type settlements, and more than 1600 villages. More than 60% of the urban population is concentrated in the capital city of Chisinau.

4.2.2 Physiography and Climate

Regionally, the territory of Moldova is part of the East European Plain. The elevation of Moldova gradually dips from an average 300 m above mean sea level (msl) in the north to less than 50 m above msl in the south near the Danube River. The land surface elevation between north and south may differ by up to 200 m. Slopes are generally gentle, which give the landscape an undulating character. Moldova is a relatively low-lying and hilly country. The average elevation is 147 m; its highest elevation is 429 m as shown in Fig. 4.24.

Moldova's climate is moderate, i.e., continental with a short mild winter and a long hot summer. Average yearly precipitation ranges from 670 mm in the north to near 450 mm in the south. In the cold season, there is a surplus of rainfall over evaporation as shown in Fig. 4.25. In the summertime, the situation is reversed, with a deficiency of rainfall. Therefore, recharge of groundwater is likely to occur in the cold season.

Precipitation varies with the season. Precipitation during the warm periods is higher than the cold periods as shown in Fig. 4.26.

The Atlantic and Arctic air currents influence Moldova's air temperature. The mean annual temperature is 7.7 °C in the north and 9.9 °C in the south. The maximum statistical air temperature in June is 22 °C (in the town of Comrat), and the minimum air temperature in January is minus 5.2 °C (in the town of Briceni). Periodically during the summer, the maximum temperature rises to 34 °C and in winter the temperature can be as low as minus 25 °C. Figure 4.27 shows Moldova's average multiannual air temperature.

4.2.3 Soil Cover

The Republic of Moldova has a large variety of fertile soils, which are considered the best in Europe. Loess and different types of clay are parent rocks for the soil stratum. On the highlands in the north and central parts of the country, gray and brown soils occur as well as grassy-podzolic soil. In the lowlands, degraded black humus (or chernozem) has accumulated. A heavy horizon of black humus covers the Belti and Budjac steppes. In the river valleys, alluvial valley soils cover the slopes and flat territories.

Agriculture is a common use of Moldova's. Erosion has already damaged about 30% of the agricultural lands in the Republic. The average thickness of the soil is between 1.2 and 2.2 m. The soil stratum is the upper part of the unsaturated zone, which typically ranges from 0.5 to 10 m in thickness and can even be as much as 10–15 m.

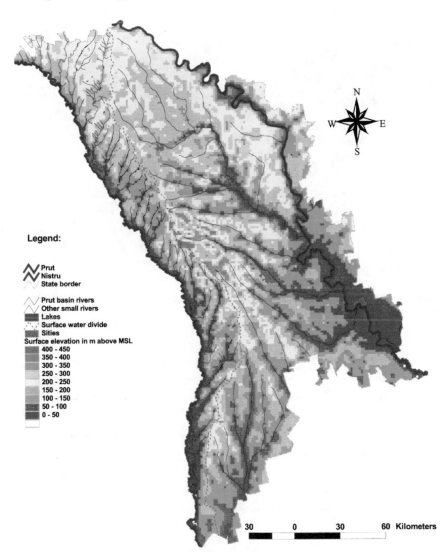

Fig. 4.24 Relief and hydrographical network of the Republic of Moldova (surface elevation is in meters)

4.2.4   Surface Water

Moldova is part of the Black Sea watershed. A relatively dense network of streams, gullies, brooks, and small rivers drains the country. The latter mainly flow toward the south and southwest in the Prut River basin and toward the southwest direction in the Nistru River basin. The general topographic layout can be seen in Figs. 4.24 and 4.28.

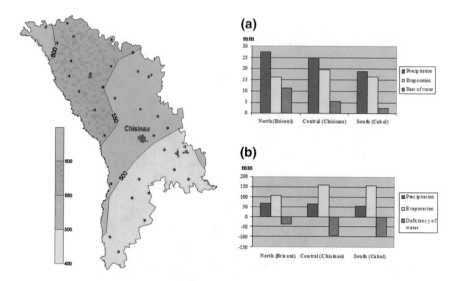

Fig. 4.25 Atmospheric precipitation and evaporation for the Republic of Moldova (**a** are data for January and **b** for July)

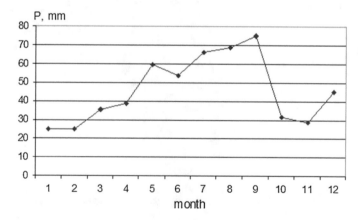

Fig. 4.26 Multiannual atmospheric precipitations, Chisinau city, Moldova

The Prut basin in the west covers roughly one-fourth of the country. Most of the thirty tributaries that drain this part of the country are branches of the Prut. The drainage system of the Nistru basin is different consisting of a few rather large tributaries, each with a dense network of branches draining large parts of central Moldova. The Nistru valley itself is rather narrow with only a few small tributaries on Moldovan side. The River Reut is by far the biggest Nistru tributary in Moldova and drains the north central part of the country. The Reut flows northeast from Balti collecting water from northern tributaries. The Reut turns southeast near Floresti and joins the Nistru some 100 km further south at Dubasari, further collecting water

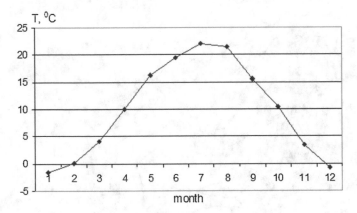

Fig. 4.27 Multiannual air temperatures, Chisinau city, Moldova

from many branches. Other large tributaries of the Nistru cross nearly the full width of the country and include the Bic which flows through Chisinau, and the Botna, which is south of the Bic. Rivers that flow into the Danube River or that flow directly into the Black Sea drain the southeastern part of the country (roughly 100 km).

Moldovan rivers have different functions. Big rivers, like Prut and Nistru, have the following uses/functions: ecosystem, fish, recreation, drinking water, irrigation, industry, safety, hydropower, and tourist transportation. Small rivers serve mainly ecosystem, recreation, and irrigation purposes.

State organizations investigate the hydrology and hydrochemistry of surface water in Moldova. These organizations monitor water quality monthly. The Hydrometeo Service (HMS), the National Centre of Preventive Medicine (CHE), the State Water Consortium "Apele Moldovei" (AM), and the State Concern "Apa-Canal" (AC) monitor water quarterly. Table 4.11 lists the parameters that each of these organizations monitors.

Local geology, neighboring countries, and local anthropogenic activities influence surface water quality. Microbiological pollution is most often detected in the small rivers. Trace elements, as well as pesticide remains, are detected in both water and sediments. Water temperature varies over time. Figures 4.29, 4.30, and 4.31 show the variability of chemical composition and temperature.

Surface water is hydraulically connected with groundwater. Shallow aquifers are discharged in the river valleys. Dip groundwater both discharges and recharges into rivers. Due to high rates of aquifer exploitation, the hydrologic regime of the Reut River in north Moldova has radically changed. In the 1950s, this river was large and only exploited locally. Now, the river is little more than a creek whose level depends on the season. Since 1991, there has been a reduction in surface water use. Pumping now averages 1/10 of what it was previously. This is due to the reorientation of Moldova's economy and agricultural production. During the Soviet period, Moldova specialized in agriculture with intense irrigation.

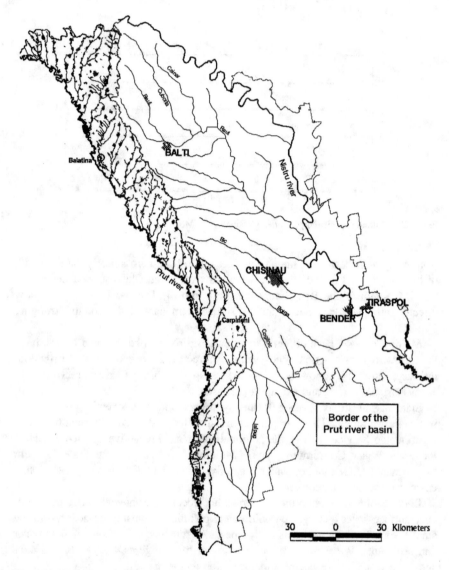

Fig. 4.28 Hydrological network of the Republic of Moldova

4.2.5   Geology and Hydrogeology

Over the past 100 years, numerous investigators have contributed to the basic knowledge of the Moldovan aquifer systems. Several investigators even published general information about the geology and hydrogeology of Moldova as early as the middle of the nineteenth century (Strucov 1852; Sinthov 1882). Between 1918 and 1945, most investigations dealt with establishing the local potential of aquifers for

Table 4.11 List of water quality parameters measured by the agencies (Tacis 2000)

Nr	Parameter		HMS	CHE	AC	DEA
1	Flow		+	*	*	*
2	Color		+	+	+	+
3	Odor		+	+	+	+
4	ph		+	+	+	+
5	Water temperature	T	+	*	+	*
6	Suspended solids		+	*	*	+
7	Transparency		+	*	+	*
8	Turbidity		*	+	+	*
9	Biological oxygen demand	BOD5	+	+	*	+
10	Chemical oxygen demand	COD	+	+	*	+
11	Oxygen (dissolved)	O_2	+	+	*	+
12	Oxygen saturation		+	*	*	*
13	Alkalinity		+	*	*	*
14	Calcium	Ca	+	+	*	*
15	Hardness		+	+	*	*
16	Magnesium	Mg	+	+	*	*
17	Chloride	Cl	+	+	*	*
18	Sulfates	SO_4	+	+	+	*
19	Total dissolved solids/salts	TDS	+	+	*	+
20	Ammonium	NH_4	+	+	*	+
21	Nitrate	NO_3	+	+	*	+
22	Nitrite	NO_2	+	+	*	+
23	Ortho-phosphate	o-PO_4	+	+	*	*
24	Total phosphorus	P-tot	+	*	*	*
25	Detergents		+	+	*	+
26	Mineral oil		+	+	*	+
27	Phenols		+	+	*	+
28	Arsenic	As	*	+	*	*
29	Fluoride	F	*	+	+	*
30	Iron	Fe	+	+	*	*
31	Manganese	Mn	*	+	*	*
32	Potassium	K	+	+	*	*
33	Sodium	Na	+	+	*	*
34	Cadmium	Cd	*	+	*	*
35	Chromium	Cr	*	+	*	*
36	Copper	Cu	+	+	*	*
37	Lead	Pb	*	+	*	*
38	Nickel	Ni	*	+	*	*
39	Zinc	Zn	+	+	*	*
40	DDT		+	+	*	*

(continued)

Table 4.11 (continued)

Nr	Parameter		HMS	CHE	AC	DEA
41	Dieldrin		*	+	*	*
42	Endrin		*	+	*	*
43	Atrazin		+	+	*	*
44	Simazin		+	+	*	*
45	alfa-hexachlorocyclohexane	a-HCCH	+	*	*	*
46	gamma-hexachlorocyclohexane	c-HCCH	+	+	*	*
47	Escherichia coli		+	+	*	*
48	fecal coliform germs		+	+	*	*
49	Salmonella		*	+	*	*
50	Enteroviruses		*	+	*	*
51	Rotaviruses		*	+	*	*
52	Viable helminti eggs		*	+	*	*
53	Enteric bacteriofagi		*	+	*	*
54	Total coliforms		*	+	*	*
55	Fecal streptococci		*	+	*	*
56	Phytoplankton		+	*	*	*
57	Zooplankton		+	*	*	*
58	Zoobenthos		+	*	*	*
59	Periphyton		+	*	*	*
60	Saprophytic bacteria		+	*	*	*

Remark *not used

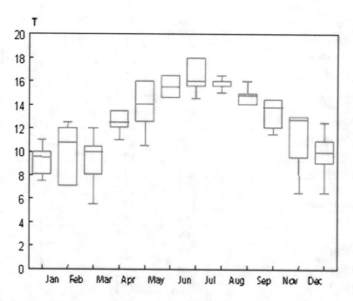

Fig. 4.29 Variance of temperature (°C), Prut River (Tacis 2000)

Fig. 4.30 Variance of ammonia, Prut River (Tacis 2000)

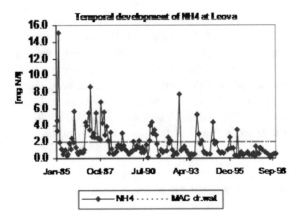

Fig. 4.31 Variance of biological oxygen demand, Prut River (Tacis 2000)

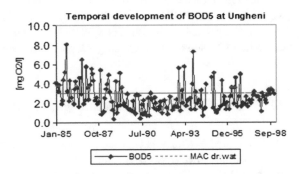

water supply (Lange 1915; Gaponov 1928; Lickov and Licistkii 1936; Macov 1940). Scholars studied only Quaternary and Neogen formations in that period. Since 1945, the interest in oil and natural gas, as well as the development of modern agricultural techniques, has stimulated new large- and small-scale geological and hydrogeological investigations. Assovskii (1954) evaluated the regional quantity and quality of groundwater. Frolov (1961) studied thermal groundwater and hydrogeologic zonation. Malevanii (1948), Vzunzdaev (1958), and Vzunzdaev (1965) examined earlier studies and gave their view of the hydrogeologic scheme of aquifers and separate layers in Moldova. Zelenin (1974) studied the hydrodynamic parameters of the main aquifers. Moraru (2009) evaluated the hydrogeochemistry of the region.

An important development was the creation of the Hydrogeological Station of Moldova in 1960. The main tasks of this station were to investigate (a) the regional groundwater regime, (b) the groundwater regime related to abstraction, and (c) the phreatic water regime of irrigated lands. The station has drilled many boreholes for these purposes, many of which were selected for the State-monitoring network, which is still active. Geologists and hydrogeologists from the State Association of Geology have long presented data related to various aspects of Moldova's hydrogeology.

Stasev (1961) was the first to calculate the potential of Moldova's groundwater resources. New data and economic requirements stimulated Saraevskii (1983) to recalculate available volumes. He estimated the available resources to be 2.3 mln m³/day. Unfortunately, 50% of this volume does not meet drinking water standards. In the field of hydrochemistry, Petracov (1972) studied fluoride in groundwater. Bondaruc (1981) investigated the distribution of nitrate pollution in natural water rocks and the relation between nitrogen compounds and land use. Zelenin (1974) studied the distribution of hydrodynamic parameters of aquifers. Moraru (1997, 2002), Moraru et al. (1985, 1990), and Melian et al. (1999) studied the hydrodynamics of the groundwater and the hydrogeochemistry.

4.2.6   General Geology

Drumea (1961), Danich and Sobetskii (1964), Bobrinsky et al. (1965), Polev and Negadaev-Nikonov (1965), Tacis (2000), and Moraru (2009) describe the primary data for the present study. Figures 4.32 and 4.33 present a geological map of the Republic of Moldova and regional geological cross sections.

4.2.7   Pre-Cambrian Crystalline Basement

The crystalline basement is exposed only in the northeast of the country (e.g., in a quarry near the town of Soroca). The exposed basement rocks typically consist of a variety of high-grade metamorphics from light-colored gneisses and schists with a sedimentary origin to very dark-colored metamorphosed (basic) igneous rocks (amphibolites and norites). Many secondary and coarse-grained light-colored veins seem to accompany the complex. The rocks, subjected to several orogenies, are tightly folded and exhibit a joint and fracture pattern from intense crustal stresses. The top of the crystalline basement is clearly visible in the quarry as a subhorizontal flat surface. No remnants were observed of a former paleosol, although a faint coloring in some places may have originated from chemical weathering during former soil formation. Silurian strata overlie the top of the basement.

Paleozoic

Only in the southern part of Moldova in the By Dobrojian depression are Ordovician, Devonian, Carboniferous, and Permian deposits supposed to be present. Very few studies have been performed on these deposits, and only general concepts are available.

Silurian Strata

Shale that grades into greenish-gray dense quartzite and shows a spectacular angular nonconformity lies over a crystalline basement. The quartzite is cm- to dm-bedded and mined as building stones. It is said that these strata may have a

GEOLOGICAL MAP

Fig. 4.32 Geological map of the Republic of Moldova and neighboring countries (Federal Geological Institute of Germany 1995) (scale 1:1,000,000 in meters, legend corresponds to international indexes)

Silurian age of 408–438 Ma years. This implies that early Paleozoic strata are missing. The top of the basement clearly indicates a widespread erosion surface.

The thickness of the Silurian strata exposed in the quarry is 10–20 m. Pleistocene deposits of the River Nistru lie directly over the quartzite. Along the flanks of the Nistru valley, distinct white-colored Cretaceous rocks appear to lie

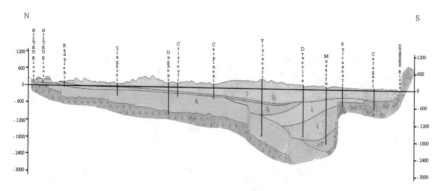

Fig. 4.33 Geological cross section across the territory of the Republic of Moldova (Drumea 1961) (legend corresponds to international indexes)

directly over the quartzite. The thickness is estimated on the order of tens of meters. It is well-known that the presence of Silurian rocks in the rest of Moldova covers the basement. Exploratory drillings have reached rocks of the Silurian age. The regional thickness of the Silurian is between several meters and 930 m and increases in the south. Ludlovian age characteristics appear only in southern Moldova. Llandoverian and Wenlockian age characteristics appear on all of the territory of the country.

Mesozoic
Jurassic Strata

These deposits are greatly developed within the By Dobrojian depression in southern Moldova. Mid- and Late Jurassic epochs are barely detectable. The authentically Middle Epoch includes the Bajocian, Bathonian, and Callovian ages. Sandstone and clay stone are the main rocks. Their total thickness averages about 1100 m. Late Jurassic includes the Oxfordian, Kimeridgian, and Tithonian ages. Arieties of colored clay, limestone with minor beds of gypsum, and anhydrite were deposited during these ages. Their total thickness is between 70 and 1330 m.

Cretaceous Strata

In the valley of the Nistru and its tributaries, Cretaceous rocks are widely exposed. They consist of the typical chalk facies, being fine-grained embedded or massive dense organic limestone. The limestone is friable, was built up by the calcispheres of foraminifera, and has a primary porosity. Many horizons of chert are recognizable. Marls (calcareous clays) can also be found as intercalations in the chalk facies. Neocomian and Albian deposits are found only in southern Moldova where their thickness is estimated to be a maximum of 440 m. Upper Cretaceous (Cenomanian, Turonian, Coniacian, and Santonian) is characteristic of all of the country's territory. The maximum thickness of the Upper Cretaceous strata is about 70 m.

Genozoic

Both the Paleogene and Neogene present the Tertiary period. The Neogene is the more important and characteristic of all of the territory of Moldova. Paleogene deposits have secondary significance, and they are composed mainly of different types of clays.

Neogene Strata

The Tortonian, Sarmatian, Meotician, and Pontician ages developed on the territory of Moldova. Tortonian deposits were discovered in many wells in the territory of Moldova. These deposits are present not only in the north, but also in the west, southwest, and southeast part of the country where intensive regression–transgression of the sea destroyed the stratum. Rocks contain many remains of the flora and fauna. Limestone and hard clay are the predominant rocks, and their thickness varies between 1 and 20 m.

The Sarmatian age stratigraphically divides into three horizons: Lower, Middle, and Upper Sarmatian. Generally, Sarmatian deposits are found everywhere. Hard limestone, sand, and clay dip toward the south and increase in thickness from several tens of meters to several hundreds. Near the southern part of the valley of the Prut River, Lower Sarmatian reef uplands formed. In the center of Moldova is the belt of the Middle Sarmatian reef. Many quarries exploit the reef limestone. In the southern part of Moldova, Sarmatian deposits outcrop.

Pontician and Meotician deposits overlie Sarmatian age and form heavy horizons in the southern part of the country. Clay, fine-grained sand, and local remains of limestone join in one unique stratum. The total thickness of the sediments is about 100 m. These deposits outcrop in many river valleys and lowland places.

Quaternary deposits

The Quaternary period is represented both by Pleistocene and Holocene epochs. Alluvial, deluvial, proluvial, eolian, and lacustrine deposits formed on top of Neogen strata. Sandwiches of clay, sand, gravel, silt, and paleosoil are the most characteristic rocks. The deposits are 1–25 m thick, sometimes more.

4.2.8 Hydrogeological Conditions

On a national scale, the hydrogeological system of Moldova consists of a sequence of "aquifers," locally indicated as water-bearing horizons, and "aquitards," or less pervious layers. From the surface downward, 17 aquifer complexes are distinguishable. Only the upper aquifers have freshwater. The lower geological formations are filled with brackish and saline water and are not subject to groundwater exploitation. Table 4.12 is a chart of major (freshwater) aquifers and aquitards, with their geographic coverage. Figures 4.34 and 4.35 show representative hydrogeological cross sections.

Table 4.12 Aquifers and aquitards distinguished in Moldova

Geological formation	Hydrogeological interpretation	Coverage of Moldova		
		South	Middle	North
Q	Local aquifers	X	X	X
N2P	Aquifer	X	–	–
N1S3 + N1 m	Aquifer	X	–	–
N1S3	Aquitard	X	X	–
N1S2	Aquitard	X	X	X
N1S2—sand	Aquifer	X	–	–
N1S2	Aquitard	X	X	X
N1S2—limestone	Aquifer	X	X	–
N1S2—clays	Aquitard	X	X	X
N1S1	Aquifer	X	X	X
K2S2 + N1b	Aquitard	X	X	X
K2S1 + S	Aquifer	X*	X*	X

*Aquifer is present, but filled with brackish groundwater

The classification "aquifers," as opposed to "aquitards," should be considered as indicative of the fact that aquifers are more pervious than aquitards. The interfaces between the aquifers and aquitards (less pervious layers) are often not very sharp. Local patches of Quaternary aquifers are unconfined. Aquifers N2P and N1S3 + N1m are either unconfined or semi-confined, depending on the region.

Over time, groundwater levels do not vary much. Variations in groundwater levels in the deep aquifers do not show a direct relation to variations in precipitation. Even in wells of shallow aquifers, deeper than a few meters, seasonal fluctuations may have been smoothed out as the result of very low hydraulic conductivity of the upper layers. The large-scale effects of withdrawal clearly appear by comparing a recent map and a map from 1960. However, development of decreasing levels is not easy to trace, at least not visually.

Groundwater abstraction in Moldova has been complicated and is becoming increasingly obscure. The first data set from 160 wells has been used to analyze the relation between withdrawal from deep wells and the hydraulic heads in observation wells. A cursory inspection of the graphs of withdrawal and the measured heads does not show any obvious relationship. The relation is either complex because of superimposed effects or absent such effects because of distances that are too large between the production and monitoring wells.

The relationship between production of larger well fields and hydraulic heads in their surroundings was also analyzed. There are little relevant data regarding the water levels in and near the well fields because of the lack of monitoring wells. Water level data are occasionally (once a year or less frequently) collected from production wells after they have stopped. This practice provided solid evidence of the relation between the depth of the depression cone and the production of the well field. Figure 4.37 consists of maps showing regional cones of depression.

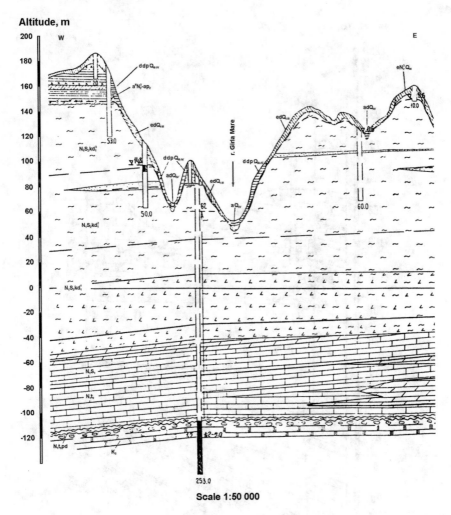

Fig. 4.34 Hydrogeological cross section, Central part of the Republic of Moldova (legend—see Fig. 4.36)

The undulating topography of Moldova and the porous nature of the Miocene sediments constitute an ideal framework for regional and local groundwater flow systems. In local or shallow systems, flow lines connect a topographic high that acts as a recharge/infiltration area with the immediately adjacent topographic low and the corresponding discharge/exfiltration area. In the case of local flow systems, the in- and exfiltration areas are juxtaposed, and flow takes place via phreatic groundwater which implies that no other flow systems are positioned on top of the system. These local or shallow groundwater flow systems are found in the fluvial dissected parts of Moldova; thus, there are hills between the streams. The infiltration or recharge areas are found on top of the hills, and discharge or exfiltration

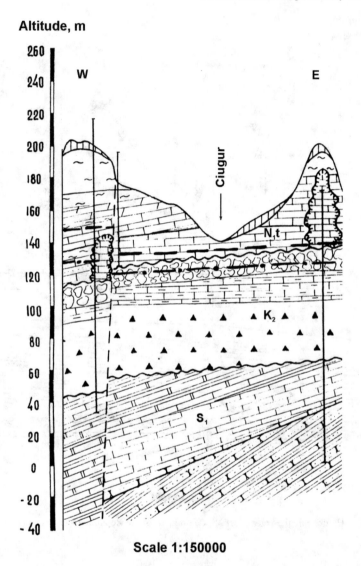

Scale 1:150000

Fig. 4.35 Hydrogeological cross section, South part of the Republic of Moldova (legend—see Fig. 4.36)

occurs in the adjacent streams. The travel times of groundwater flow in these systems may be on the order of years to tens of years. The majority of dug wells tap the water from these superficial groundwater systems.

Regional flow systems top the hierarchical organization and are the highest level of scale. All other flow systems (local and subregional) are within the regional one. The recharge areas from the highest morphological plateau drive these regional systems. Discharge occurs along the valleys of the rivers Prut, Nistru, and Reut.

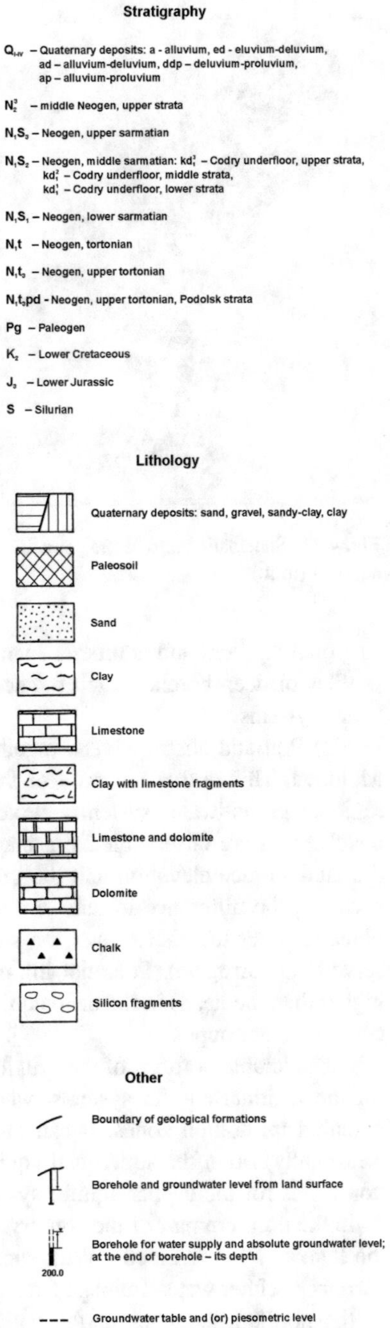

Fig. 4.36 Legend to geological cross sections

Stratigraphy

Q_{uv} – Quaternary deposits: a - alluvium, ed - eluvium-deluvium, ad – alluvium-deluvium, ddp – deluvium-proluvium, ap – alluvium-proluvium

N_2^2 – middle Neogen, upper strata

N,S_2 – Neogen, upper sarmatian

N,S_2 – Neogen, middle sarmatian: kd_1^3 – Codry underfloor, upper strata, kd_1^2 – Codry underfloor, middle strata, kd_1^1 – Codry underfloor, lower strata

N,S_1 – Neogen, lower sarmatian

N,t – Neogen, tortonian

N,t_2 – Neogen, upper tortonian

N,t_2pd - Neogen, upper tortonian, Podolsk strata

Pg – Paleogen

K_2 – Lower Cretaceous

J_2 – Lower Jurassic

S – Silurian

Lithology

Quaternary deposits: sand, gravel, sandy-clay, clay

Paleosoil

Sand

Clay

Limestone

Clay with limestone fragments

Limestone and dolomite

Dolomite

Chalk

Silicon fragments

Other

Boundary of geological formations

Borehole and groundwater level from land surface

Borehole for water supply and absolute groundwater level; at the end of borehole – its depth

200.0

Groundwater table and (or) piesometric level

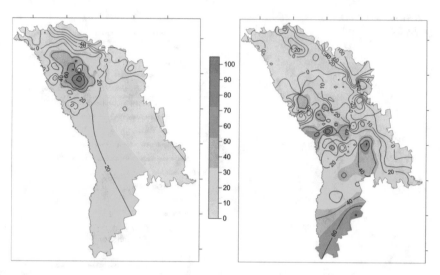

Fig. 4.37 Schematic maps of the potentiometric surface of Cretaceous (left) and Sarmatian (right) aquifers (in m)

Regional aquifers and aquitards span large parts of the country. The outcrops and profiles of deep boreholes tell the scope of these large and deep regional groundwater systems.

The Prut and Nistru Rivers, together with their tributaries, drain a large part of Moldova. This is true not only of the surface water system but also of a part of regional groundwater systems. "Potential infiltration" zones and "potential drainage" zones are observable from maps that show the difference in levels between the land surface elevation and the groundwater head in the deep aquifers. (To be accurate, the difference in land surface elevation is not significant, but rather the phreatic water table. However, on such a large scale, land surface elevation may serve as a surrogate). Potential infiltration zones occur where the land surface is higher than the heads in the deep aquifers; potential drainage zones occur where the conditions are opposite.

Considerable portions of the Prut and Nistru valleys are potential drainage zones for the regional aquifer systems, whereas the areas between the rivers constitute potential infiltration zones. It can also be inferred that many small streams may (seasonally) drain the superficial aquifers and simultaneously be potential infiltration zones for the deeper aquifer systems.

In the northern part of the country, some of the upper aquifer systems surface in the flanks of the riverbeds. With such free flow surfaces, there are no significant barriers to either water drainage from groundwater systems or infiltration into them. In the northern stretches of the valleys of the Prut and Nistru Rivers, drainage appears to be the dominant process. Toward the south, the aquifer systems incline to greater depths. Especially in the central and southern part of the country, thick and dense aquitards cover the aquifers, thereby preventing open connections with the rivers. The hydraulic resistance of these aquitards may become so high that the aquifer systems beneath can be considered semi-confined to fully confined. Positive or negative groundwater gradients remain determinative of the direction of vertical groundwater flow, but the fluxes may become extremely poor. They may only be significant in a time frame of ten thousands to millions of years. Many deep wells penetrate these deep aquifer systems. Their production and drawdown may vary considerably. Local wells of industries or former state farms with a limited production will probably not cause large depression cones. However, it appears that some large well fields installed for the water supply of cities create considerable depressions spread over relatively large distances. More than 6600 wells were drilled for groundwater extraction. In 2012, only 3100 wells were in production. Approximately 250,000 wells located mainly in rural areas exploit the shallow aquifer. The total reserves of dip aquifers amount to 3,173,000 m^3/day. The reserves of shallow aquifers have not been determined. For the territory of the Republic of Moldova, fresh, thermal, technical, mineral, and industrial (containing I, Br, B, and He) groundwater are typical.

4.2.9 Tectonics and Paleohydrogeology

Moldova sits atop a stable Pre-Cambrian (much older than 560 Ma years) crystalline shield, often denoted as the European Platform. It is known locally as the Ukrainian crystalline shield. This very old stable platform in fact represents the roots of mountain ranges formed during the many orogenic cycles in Archeozoic and Proterozoic times (2300–560 Mya). Long periods of erosion leveled the mountains, which eventually resulted in a relatively flat platform that may have been a landmass for a long period prior to the Paleozoicum (starting at 560 Mya). It follows that the crystalline platform consists of highly metamorphic rocks that may have been subjected to several orogenic cycles. The literature on the geology of Moldova suggests that this crystalline platform has been reached in many exploration drillings throughout the country.

The rocks overlying the stable platform are nearly formless with regional sub-horizontal strata. Much of the stratigraphy overlying the crystalline basement is absent because of the lack of deposits or erosion afterward. These stratigraphical gaps in sedimentation indicate that the area may have been a landmass for much of the time. Figure 4.38 shows that the territory of Moldova is divided in several tectonic units.

An important geological event that has influenced the geological development of Moldova is the Alpine Orogeny (Tacis 2000). Indirectly and in a geological sense, this event created most of Moldova as well as the present-day groundwater. Deep crustal processes are related to the Alpine Orogeny which in fact represents the collision of the African and European continents and have influenced the crystalline basement even though it is located outside of the Alpine orogenic belt. The Carpathian Mountains are west of Moldova and form a peculiar curved belt in

Fig. 4.38 Tectonic scheme of the Republic of Moldova (Polev and Negadaev–Nikonov 1969)

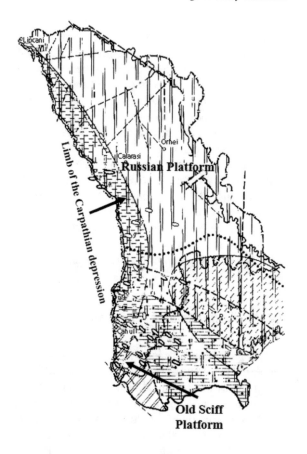

Hungary and Romania. This mountain belt consists of eastward upthrusted strata (an even nappe formation) onto the stable crystalline basement that extends underneath Moldova. This has caused down-buckling of the crystalline basement under the enormous load of the upthrusted strata. This flexural loading and down-buckling of the basement, together with other crystal processes related to continental collision, caused subsidence of the area below sea level; Miocene sea levels were much higher than the present sea level. This occurred during the Miocene period, from 5 to 25 Mya. Additionally, the upthrusting of strata in the East Carpathians caused a huge amount of sediments to be deposited in the newly created basins east of the Carpathians, i.e., in Moldova and surroundings. This explains the very thick pile (some hundreds of meters) of Miocene marine clayey rocks in Moldova, which now form the major water-bearing strata. The Miocene pile shows a trend from open marine conditions in the lower parts to near-coastal (lagoons, deltas and swamps) settings in the upper part, indicating the filling of the Alpine basins.

The Miocene sediment pile is of paramount importance to groundwater and groundwater flow systems. The underlying rocks of the Paleozoic and Mesozoic ages (sandwiched between the crystalline basement and the Miocene rocks) are much less important hydrogeologically. They are of interest for oil and gas exploration except for the northern part of Moldova where the top of the crystalline basement becomes shallow and is tapped by deep wells. The Paleozoic and Mesozoic rocks are exposed in the valley of the River Nistru.

Present-day Moldova is built up by the clayey and sandy marine rocks of the Miocene age. These rocks are exposed throughout the country; locally there are some thin veneers of Quaternary alluvial sediments. Eolian loess deposits from the Pleistocene glacial periods are widespread in Moldova, giving rise to fertile black-colored loamy chernozem soils. In terms of Alpine Geology, these Middle to Late Miocene rocks (10–15 Mya) represent the so-called Molasse deposits of the Eastern Carpathian mountain ranges.

The Miocene strata and their geological development have also dominated geomorphological development during the Quaternary. After Miocene sedimentation ceased and the area emerged from the sea because of the global drop in sea level, a flat landmass or platform was created. Initially the Prut and Nistru Rivers may have flown through this vast plain as broad disorganized meandering streams. One may even speculate that they were one river. After a further drop in sea level and/or a combination of tectonic uplifts after the Alpine basin formation, the rivers must have started to incise the plain and to create broad valleys that finally confined the rivers to their present-day relatively narrow valleys. Especially along the Prut, one may notice the remnants of the former broad valleys as terraces, "hanging" shoulders and breakthrough gorges. What remains is a kind of staircase morphology with what was probably the original plain in the centre of the country bounded by former broad terraces and valleys in directions descending to both the present-day river courses. A secondary drainage pattern has developed with many tributaries draining into the Prut and Nistru. The end result is a set of plains incised by relatively steep young valleys which extend over great distances. Especially along the valleys, spectacular mass movements and earth slides occur continuously on a scale ranging from meters to several kilometers in these plastic clayey Miocene sediments. Large-scale earth slides tend to push up series of peculiar elongated hills along the front of the slide, giving rise to closed depressions filled with lakes and wetlands.

In the northern half of the country, clayey and marl marine sediments containing a lot of calcium carbonate built up Miocene rocks. When exposed, the sediments appear as monotonous light brown-colored clays, marls and mud stones, often without any bedding plane or sedimentary structures. There are many horizons of local reef limestone found enveloped in the clayey monotonous sediments. These reef lenses represent periods of interrupted sedimentation caused by an insufficient supply from Carpathian sources coupled with a shallower environment (sedimentation outpacing basin subsidence). One often finds former beaches along the reef lenses consisting of carbonate sands. Once sedimentation was reestablished, the clayey sediments engulfed and finally buried the reefs thereby causing the reef-building organisms to perish. Upon inference from exposures, the sedimentary environment is probably an open marine shelf environment outside the influence of waves (low energy). Some exposures show horizons (1–3 m) of cross-bedded sands, indicating high-energy near-coastal (littoral) environments.

Hydrogeologically, the overall predominance of clayey sediments is very important to subordinate discontinuous limestone horizons (5–15 m) and sand layers. The sediments nearly always contain carbonate but macrofossils are rare. Another important point is that these marine sediments, in particular the clayey strata, contain a lot of entrapped seawater from the time of deposit which due to the geological history and development of a drainage pattern has not been fully flushed.

In the southern half of the country, Miocene rocks have been deposited in a different sedimentary environment. The rocks slope gently southward and the younger part is exposed in the south. The rocks contain many more sand layers. Many exposures show sandy deposits with cross bedding giving the impression of littoral (near-coastal) environments well within the wave base. Limestone is a minor phenomenon. Also, there are delta environments with sandy erosion gullies in the more silty pro-delta sediments. Allochthonous gravels have been found in the delta gullies and even in beach deposits. It was said that further to the southeast the Germans mined brown coal during the Second World War indicating upper deltaic and coastal swamp-like environments. Apparently, the sedimentation outpaced the subsidence and the basin gradually filled, thereby creating terrestrial conditions. The high amount of sandy deposits is important to groundwater flow. Another important point may be the presence of organic materials from coastal swamp environments. Organic materials are strong chemical reducing agents when in contact with groundwater. As a result of geological development, different tectonic faults were formed on the territory of the Republic of Moldova as presented in Fig. 4.39.

Primarily, all aquifers in Moldova's cross section have sedimentation of marine origin. The oldest complex of aquifers belongs to Paleozoic permeable deposits. Water chemistry was predominately chloride and sodium. As the sea regressed, the northern part of the country was exposed to surface conditions while the southern part remained under a shallow sea. Infiltration of atmospheric precipitations and surface waters extruded saline water in the middle of Moldova where the Silurian horizon dips sharply under Mesozoic formations.

Approximately 70–250 Mya, Mesozoic aquifers filled with seawater. The Cretaceous aquifer in the northern part of Moldova changed the water chemistry analogically to the Silurian. A Jurassic complex of aquifers formed in the By Dobrijia depression, and its geochemistry remains the same. The TDS in these aquifers are between 60 and 300 g/l with high concentrations of trace elements, namely iodine, boron, rubidium, and bromine.

Tertiary aquifers formed under the influence of the Sarmatian paleosea. Initially, water in the Neogen water-bearing rocks was middle saline corresponding to chloride sodium/calcium type. Flat sedimentation of deposits and its inclination from North to South provided good conditions to flush horizons by infiltrated surface and precipitated water. Until the Quaternary period, Neogen aquifer water

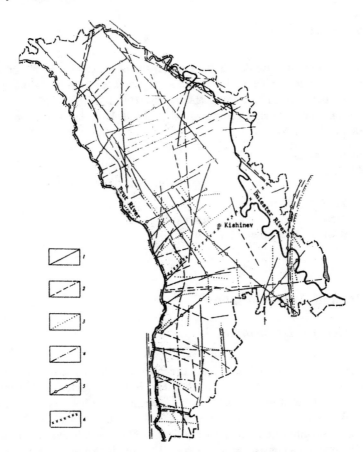

Fig. 4.39 Maps showing principal faults, Republic of Moldova (Bobrinsky et al. 1986, 1987). In the legend faults as *1* axial lines of helium anomalies, *2* gravitational field lines, *3* axial zones of mantle anomalies, *4* faults identified by drilling, *5* faults identified by geomorphological methods, and *6* Vrancea–Kanev deep mantle lineament

chemistry drastically changed. Paleogeographical investigation indicates that water of Neogen horizons was brackish. The processes of flushing and water popping continue to the present time. Isotopic data show that the youngest water in Neogen aquifers is about 50 years (Northern part of Moldova) and the oldest is approximately 5000 years (Central part of Moldova). The water type is sodium bicarbonate to sodium chloride. The Black Sea is a regional area of groundwater discharge.

Quaternary aquifers changed their chemical composition in the interval of TDS 1–6 mg/l. In the hot periods, the TDS increased and vice versa during humid periods. Generally, atmospheric precipitations and ascending artesian water infiltrations are the main factors determining the history of groundwater geochemistry. These mechanisms still function.

4.2.10   Land Use

The collection of land use data is for the purpose of understanding the impact of land use, especially agricultural and industrial practices on groundwater quality. Information is presented on a topographic map, scale 1:200,000, only for the Prut River basin. The following categories of land use have been distinguished as illustrated in Fig. 4.40:

- residential areas
- arable lands
- pastures
- perennial lands
- undeveloped lands (forests, ravines, etc.)
- surface water (lakes, ponds, etc.).

4.3   Rastatt Area, Germany

4.3.1   General Delineation

Data exposed in the subchapter 4.3 are described in detail by Eiswirth and Hotzl (1997), Eiswirth et al. (2003), Wolf et al. (2003, 2007) and generalized by authors. Rastatt area is located in the South-West of Germany and belongs to the state of Baden—Wurttemberg (Fig. 4.41). Rastatt City is the main populated location of this region. The city extension is 6.7 km from North to South and 7.9 km East to West. The total area consists about 53 km². The area is mainly flat, with only the center part consisting of any rise in altitude. The study area includes the City of Rastatt and the communities incorporated within its administrative boundaries.

4.3.2   Physiography and Climate

The topography is quite flat with the altitude of the city center being 115.5 m a.s.l., the highest altitude within the administrative boundary is 130.0 m a.s.l., and the lowest being 110.5 m a.s.l. Climate conditions are continental. The mean annual temperature is 10 °C, mean annual precipitation is between 850 and 1000 mm. Statistical atmospheric precipitation data for the Rastatt area are presented in Fig. 4.42 (Yearly trends 2014).

Air temperature is seasonally changed and is characteristic for continental conditions (Fig. 4.43). Wind speed is statistically medium (Fig. 4.44). Nevertheless, such speed and flat surface topography increase the evaporation of water from both surface water bodies and unconfined aquifers.

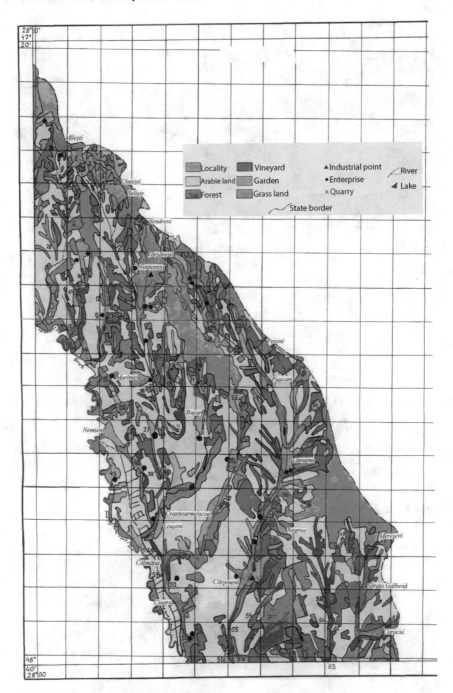

Fig. 4.40 Example of land use for the Republic of Moldova (Prut River valley) (Tacis 2000)

Fig. 4.41 Location of the study area of Rastatt. The model area is marked gray

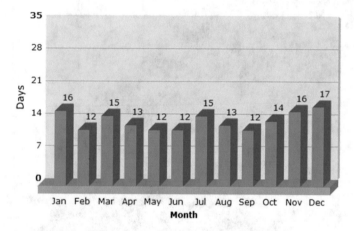

Fig. 4.42 Number of rain/drizzle days in month for Rastatt area

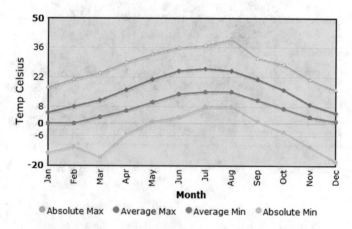

Fig. 4.43 Air temperature fluctuation in Rastatt area

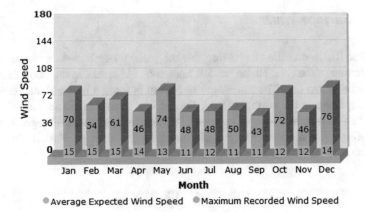

Fig. 4.44 Wind speed in Rastatt area

4.3.3   Soil Cover and Unsaturated Zone

Soil cover in the study area is used locally for domestic agriculture. Limited data are available for characteristics of soil. Thickness of the soil stratum varies between 0.2 and 1.0 m. Bellow the soil sand, sand with gravel, compressed sand and gravel are dominated in cross section. Geologically, this complex of rocks is stratified in the following units (Wolf et al. 2007). The lower terrace is morphologically clearly separated by an 8–10 m high margin in the north of Rastatt. The lower terrace is mainly set up by gravel and sands of the Würm ice age and partly covered with sand drift. In the Holocene, the Rhine River cut channels up to 20 m deep into the surface of the lower terrace. This erosive material was partly redeposited in the lowlands of the Rhine River and sporadically superposed by flood deposits, which resulted in a 1–2 m thick layer of loam. To the east of the lower terrace runs the Kinzig-Murg-Channel. This system has eroded into the sediments of the lower terrace and is partly filled with postglacial sand, clay and peat. The fluviatile depositional facies lead to a pronounced heterogeneity of the sediments on a small to medium scale, which is also visible in the depth to groundwater table maps of the local authorities. The Holocene sedimentary cover clearly exhibits the influence of the meandering rivers. Besides the common sand and gravel sediments, silty and clayey overbank fines and trench fillings are also found frequently. The topmost sediments may be grouped into (i) Holocene Floodplain Loam (Auenlehm), (ii) Holocene Floodplan Sand (Auensand), (iii) Holocene unclassified (ungegliedert), and (iv) Pleistocene Terrace Gravels (Terrassenschotter). In many places described deposits are anthropogenically refilled.

4.3.4 Surface Water

The river Rhine and the river Murg are the main water bodies in the area. The
straightening of the river Rhine has left numerous old creeks in the floodplains that
have partially run dry. The townscape of Rastatt is formed by the river Murg. The
Murg originates in the northern Black Forrest was straightened and embanked. To
the east of Kuppenheim a small channel ("Gewerbekanal") was diverged from the
river Murg, which meets Murg again in the city center of Rastatt (Fig. 4.45)
(Eiswirth et al. 2003).

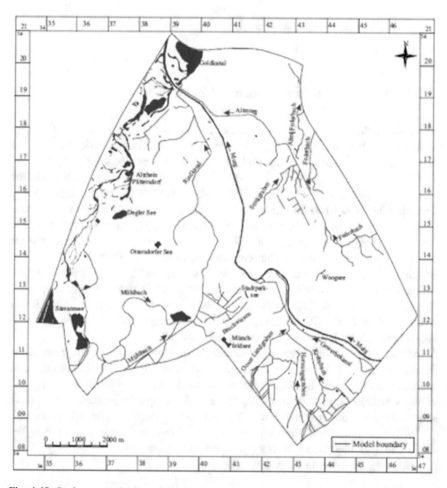

Fig. 4.45 Surface water bodies of the Rastatt area

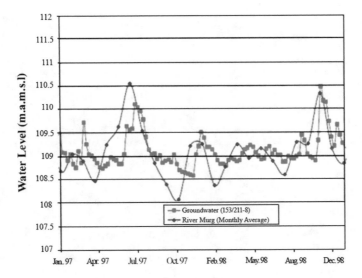

Fig. 4.46 Groundwater and surface water level correlation

Water quality is changeable in time depending of seasons and anthropogenic activity. For example, for the river Murg the following is characteristic: conductivity is fluctuating from 50.0 to 100.0 uS/cm, DOC from 2.0 to 5.0 mg/l, chloride from 4.0 to 10.0 mg/l and boron from 0.0 to 4.0 mg/l.

In the study area surface water and groundwater (unconfined aquifers) are hydraulically connected. Surface water bodies have major effects on groundwater levels (Fig. 4.46) (Eiswirth et al. 2003). Detailed investigations confirm that this connection has highly influenced groundwater quality. A sewer system is the major source of pollution.

4.3.5 Geology and Hydrogeology

From geological point of view, the investigated territory is a part of Upper Rhine graben (Figs. 4.47 and 4.48). The geology of the Upper Rhine graben in the area of Rastatt can be divided into four main tectonic units: a graben block, a downfault block, a marginal block (fore hills), and a outlier zone (basement). During the Holocene epoch, the river Rhine excavated the Lower Terrace Formations within the present Rhine depression and filled it up with recent material in various thicknesses. With generally a clearly rising slope, the Rhine depression borders the valley terrace that in this region corresponds to the uniform flat Lower Terrace

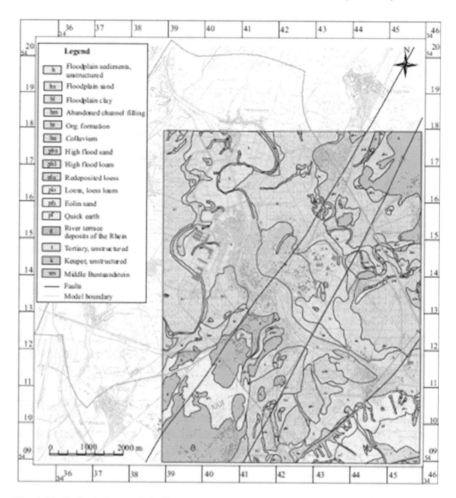

Fig. 4.47 Geological map of the Rastatt test—site, Germany (Eiswirth et al. 2003)

plain. Along the mountain border, the rivers and others coming down from the mountains created a Holocene channel. In the region of Rastatt the Pleistocene sediments are some 60 m thick with the thickness increasing continuously toward the north (Fig. 4.48). On the graben block the Upper and Middle Pleistocene gravel layer is about 33 m thick. Below follow the Lower Pleistocene Rhine sediments: the base of the Quaternary sediments is built by impermeable Pliocene sediments at an elevation of about 55 m a.s.l. (Eiswirth et al. 2003).

The hydrogeology of the area is combined in four major aquifers: Upper Gravel layer (qOKL), Middle Gravel layer (qMKL), Lower Quaternary (qA) and Pliocene (tPL) (Figs. 4.49 and 4.50—lines of the cross sections—see Fig. 4.47).

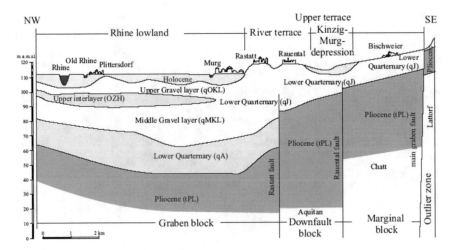

Fig. 4.48 Geological cross section (NW–SE) through the Upper Rhine valley in the area of Rastatt (Eiswirth and Hotzl 1997)

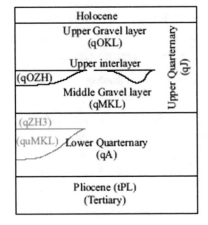

Fig. 4.49 Lithostartigraphical classification of the sediments in the Rastatt area (Watzel and Ohnemus 1997)

The hydrogeochemistry of the Rastatt area is well investigated both in spatial and vertical distribution. For a large set of ingredients, the statistical provenance was determined. The representative groundwater chemistry is summarized in Table 4.13.

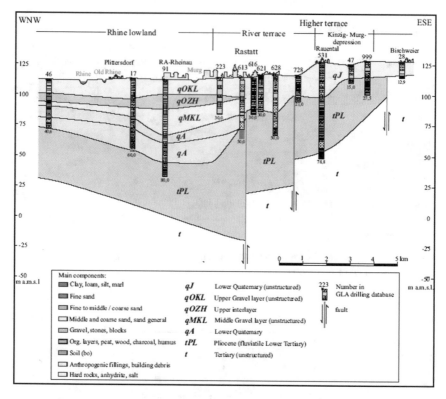

Fig. 4.50 Hydrogeological cross section of the study area (Eiswirth et al. 2003)

The predominant natural water type is HCO_3–Ca. and probably HCO_3 SO_4–NaCa is water type for anthropogenic conditions.

4.3.6 Land Use

The investigated area is very economically developed. The City of Rastatt is situated in one of the economically strongest regions of Germany. The economy is based on local medium-sized companies on the one hand and internationally active establishments of industry like the A-class plant of DaimlerChrysler, Maquet-Gettinge or Siemens on the other. Service, trade, and production can be found conterminously. The main sectors of economy are connected with agriculture and forestry, mining and quarrying of stone, food industry and tobacco processing, textile, chemical industries, metal production and processing, manufacturing of durex and plastics. The land use details are summarized in the Table 4.14.

Table 4.13 Groundwater quality of unconfined aquifer, Rastatt site, Germany (in mg/l) (Eiswirth et al. 2003)

	Minimum	Maximum	Mean $\bar{x}$
T [°C]	7.9	18.0	12.79
SEC [µS/cm]	104.0	935.0	529.08
pH	6.16	7.8	7.21
O_2	0.1	12.7	4.63
Ca^{2+}	27.3	142.0	94.54
Mg^{2+}	1.0	14.0	7.35
Na^{2+}	2.9	30.0	12.12
K^+	0.05	28.0	4.11
Fe^{3+}	0.01	18.6	0.79
Mn^{2+}	0.001	2.7	0.13
Cl^-	8.7	52.0	21.89
SO_4^{2-}	19.9	73.0	39.07
NH_4^-	0.005	2.332	0.071
NO_3^-	0.4	50.7	15.5
NO_2	0.01	1.3	0.028
HCO_3^-	73.0	406.0	269.1
o-PO_4	0.01	0.159	0.036
SiO_2	2.1	18.0	9.44
B	0.002	0.5	0.049
Al	0.001	0.11	0.0197
As	0.001	0.005	0.0032
Ba	0.023	0.3	0.0768
Ni	0.001	0.023	0.0052
Zn	0.01	0.46	0.08
F	0.02	0.45	0.108

Table 4.14 Land use data for Rastatt area (Eiswirth et al. 2003)

Land use and infrastructure	Area [ha]	Proportion of model area [%]	Proportion of administrative area [%]
Study area	10,104.5	100.0	
Administrative area	5901.0	58.4	100.0
Agricultural land	288.0	25.6	43.8
Forest area	1374.0	13.6	23.3
Water	27.0	0.3	4.6
Total of rural area	3989.0	39.5	67.6
Green space	243.0	2.4	4.1
Traffic area	15.2	1.5	2.6
Developed area	1054.0	10.4	17.9
Total of urban area	1312.2	1.4	22.2

References

Annual Data. (2004). http://madp.sws.ninc.edu/nadpdata/annualRed.Asp?Site=TN14.

Arthur, J. K. & Taylor, R. E. (1990). *Definition of the geohydrologic framework and preliminary simulation of ground-water flow in the Mississippi embayment aquifer system, Gulf Coastal Plain*. U. S. Geological Survey Water-Resources Investigation Report 86-4364, 97p.

Assovskii, G. N. (1954). *Ground water of Moldova and its practical significance for water supply* (200 p). Russia: Ph.D., Moscow State University (in Russian).

Bobrinsky, B. M. Kaptsan, V. X., & Safarov Y. I. (1965). *Paleogeography of Moldova* (122 p). Chisinau: Cartea Moldoveneasca (in Russian).

Bobrinsky, B. M., Macarescu, V. S., & Moraru, C. E. (1986). *Reflection of various tectonic violations in the helium, macro seismic and hydrogeochemical fields of Moldova*. Reports of the U.S.S.R. A. S., Vol. 288, No 5, pp. 1186–1189 (in Russian).

Bobrinsky, V. M., Macarescu, V. S., & Moraru, C. E. (1987). Tectonic factors governing the structure of the helium, macro seismic and hydrochemical regions of Moldovia. *Geotectonics, 2*(2), 150–160.

Bondaruc, N. T. (1981). Nitrate in Natural water and rocks of Moldova (187 p). *Ph.D Dissertation*. Russia: Leningrad State University (in Russian).

Boswell, E. H., Moore, G. K., MacCary, L. M., Jeffery, H. G., & et al. (1965). Water resources of the Mississippi Embayment; cretaceous aquifers in the Mississippi embayment with discussions of quality of water. *U.S. Geological Survey Professional Paper 448-C*, 37.

Boswell, E. H., Cushing, E. M., Hosman, R. L. (1968). Quaternary aquifers in the Mississippi embayment." *U.S. Geological Survey Professional Paper 448-E*, 15.

Brahana, J. V., Mesko, T. O., Busby, J. F., & Kraemer, T. F. (1985). *Ground-water quality data from the Northern Mississippi embayment; Arkansas, Missouri, Kentucky, Tennessee, and Mississippi*. U.S. Geological Survey Open-File Report 85-683, 15 p.

Brahana, J. V., Bradley, M. W., & Dolores, M. (1986). *Preliminary delineation and description of the regional aquifers of Tennessee; tertiary aquifer system*. U.S. Geological Survey Water-Resources Investigations Report 83–4011, 23 p.

Brahana, J. V., Parks, W. S., Gaydos, M. W. (1987). *Quality of water from freshwater aquifers and principal well fields in the Memphis Area, Tennessee*. U.S. Geological Survey Water-Resources Investigations Report 87–4052, 22 p.

Brahana, J. V. & Broshears, R. E. (2001). *Hydrogeology and Ground-Water Flow in the Memphis and Fort Pillow Aquifers in the Memphis area, Tennessee*." U.S. Geological Survey Water-Resources Investigation Report 89-4131, 56 p.

Coupe, R. H., Manning, M. A., Foreman, W. T., Goolsby, D. A., & Majewski, M. S. (1999). *Occurrence of pesticides in rain and air in urban and agricultural areas of Mississippi, April–September, 1995*. U.S. Geological Survey Toxic Substances Hydrology Program-Proceedings of the Technical Meeting, Charleston, South Carolina, March 8–12, 1999 (Vols. 2 and 3). U. S. Geological Survey Water-Resources Investigations Report 99-4018B.

Cox, R. T., & Van Arsdale, R. B. (2002). The Mississippi embayment, North America: A first order continental structure generated by cretaceous superplume mantle event. *Journal of Geodynamics, 34,* 163–176.

Criner, J. H., & Armstrong, C. A. (1958). Ground-water supply of the Memphis Area. *U.S. Geological Survey, Circular, 408*, 20.

Criner, J. H., Sun, P. C., Nyman, D. J. (1964). Hydrology of the aquifer system in the Memphis Area, Tennessee." *U.S. Geological Survey Water-Supply Paper* 1779–0, 54 p.

Cushing, E. M., Boswell, E. H., & Hosman, R. L. (1964). General geology of the Mississippi embayment. *U. S. Geological Survey Professional Paper* 448-B, 28 p.

Danich, M. M., & Sobetskii, V. A. (1964). *The stratigraphy of sedimentary formations of Moldova* (40 p). Chisinau: Cartea Moldoveneasca. (in Russian).

Drumea, A. V. (1961). *Tectonics of the Moldovskoi SSR* (60 p). Academy of Sciences of USSR.

Edwin A.B. and Nyman D.J. 1968. Flow Pattern and Related Chemical Quality of Ground Water in the "500-foot" Sand in the Memphis Area, Tennessee. *U.S. Geological Survey Water-Supply Paper 1853*, 27 p.

Eiswirth, M., Held, I. I., Wolf, L., & Hotzl, H. (2003). *AISUWRS work package* 1. Background study. University of Karlsruhe, Department of Applied Geology, Commissioned report 30.08.2003, 78 p.

Eiswirth, M., & Hotzl, H. (1997). The impact of leaking sewers on urban ground water. In: J. Chilton et al. (Eds.), *Ground water in the urban environment: Vol. 1. Problems, processes and management* (pp. 399–404).

Evaporation. (2014). http://www.climate.gov/tags/evapotranspiration.

Flohr, D. F., Garrett, J. W., Hamilton, J. T.,& Phillips, T. D. (2003). *Water resources data Tennessee water year 2002*. U.S. Geological Survey Water-Data Report TN-02-1, 442 p.

Frolov, N. M. (1961). Groundwater of the pre-blak see Artesian Basin. Hydrogeological Lab. F.P. Savarenskii, V.37 (in Russian).

Gaponov, E. A. (1928). List of Boreholes and Hydrogeological Map of the South-East Ukraine, Odessa (in Russian).

Gonthier, G. J. (2002). *Quality of shallow ground water in recently developed residential and commercial areas, Memphis Vicinity, Tennessee, 1997*. U. S. Geological Survey Water-Resources Investigation Report 2002-4294, 105 p.

Hardeman, W. D. (1966). *Geological map of Tennessee (west sheet)*. Tennessee: Department of Conservation, Division of Geology.

Hosman, R. L., Long, A. T., Lambert, T. W., Jeffery, H. G., & et al. (1968). Water resources of the Mississippi embayment; tertiary aquifers in the Mississippi embayment, with discussions of quality of the water. *U.S. Geological Survey Professional Paper 448-D*, 29 p.

Hosman, R. L., (1996). Regional stratigraphy and subsurface geology of cenozoic deposits, Gulf Coastal Plain, South-Central United States. *U. S. Geological Survey Professional Paper 1416-G*, 35 p.

Hutson, S. S., & Morris, A. J. (1992). *Public Water-supply systems and water use in Tennessee, 1988.*" U. S. Geological Survey Water-Resources Investigation Report 91-4195, 74 p.

Kazmann, R. L. (1944). *The water supply of the Memphis Area: A progress report*. U.S. Geological Survey, 66 p.

Kingsbury, J. A., & Parks, W. S. (1993). *Hydrogeology of the principal aquifers and relation of faults to interaquifer leakage in the Memphis Area, Tennessee*. U.S. Geological Survey Water-Resources Investigations Report 93-4075, 18 p.

Kleiss, B. A., Coupe, R. H., Gonthier, G. J., & Justus, B. G. (2000). Water quality in the Mississippi embayment, Mississippi, Louisiana, Arkansas, Missouri, Tennessee, and Kentucky. *U.S. Geological Survey Circular 1208*, 35 p.

Lange, O. C. (1915). About hydrogeological investigations of the territory of Basarabia. Agriculture of Bassarabia, No 9 (in Russian), 10 p.

Larsen, D., Gentry, R. W., Ivey, S., Solomon, D. K., & Harris, J. (2002). Ground water leakage through a confining unit beneath a municipal well field, Memphis, Tennessee, USA. In H. D. Schulz & A. Hadeler (Eds.), *Geochemical processes in soil and groundwater* (pp. 51–64). Berlin: Wiley.

Lickov, B. A., & Licistkii, V. I. (1936). Map of the hydrogeological regions in Ukraine (in Ukrainian).

Macov, K. I. (1940). *Groundwater of the Pre-Blak see depression*. Gosgeolizdat (410 p) (in Russian).

Malevanii, E. T. (1948). *Essay of the history of the hydrogeologics investigations in Moldova and Izmail Region* (15 p) (in Russian).

McMaster, B. W., Parks, W. S., Scott, W. (1988). *Concentrations of selected trace inorganic constituents and synthetic organic compounds in the water-table aquifers in the Memphis Area, Tennessee*. U. S. Geological Survey Open-File Report 88-485, 23 p.

Melian, R., Myrlean, N., Gurev, A., Moraru, C., & Radstake, F. (1999). Groundwater quality and rural drinking water supplies in the Republic of Moldova. *Hydrogeology Journal, 7*(3), 188–196.

Moore, G. K., & Brown, D. L. (1969). *Stratigraphy of the fort pillow test well, Lauderdale County, Tennessee*. Tennessee Division of Geology, Report of Investigation No 26. Boswell.

Moraru, C. E. (1997). Helium geology of the Republic of Moldova. *Intelectus, 1997,* 18–25. (in Romanian).

Moraru, C. E. (2002). Contribution to the study and practical use of ground water of Moldova. In: *Studies and practical reports related to water resources management in a vulnerable environment conditions* (pp. 32–40) (in Romanian).

Moraru, C. E. (2009). *Gidrogeohimia podzemnyh vod zony activnogo vodoobmena krainego Iugo-Zapada Vostocno—Evropeiskoi platformy* (210 p). Chisinau: Elena V.I.

Moraru, C. E., Bobrinsky, V. M., & Milcova, L. N. (1985). *Investigation of the chemical composition of Moldova groundwater with paleohydrogeological, hydrodynamic and geochemical consideration*. Report IGG, Chisinau, 460 p (in Russian).

Moraru, C. E., Burdaev, V. P., & Negrutsa, P. N. (1990). Classification and evaluation of hydrogeochemical facies using the cluster analysis. Deposited with VINITI, 1990, No 6497-V90, Moscov, 15 p.

Myrlean, N. F., Moraru, C. E. & Nastas, G. E. (1992). *The ecological and geochemical atlas of the city of Chisinau* (191 p). Chisinau: Stiinta (in Russian).

Neely, B. L., Jr. (1984). *Flood frequency and storm runoff of urban areas of Memphis and Shelby County, Tennessee.* U.S. Geological Survey Water-Resources Investigations Report 844110.

Nyman D.J. 1965. *Predicted hydrologic effects of pumping from the Lichterman well field in the Memphis area, Tennessee."* U.S. Geological Survey Water-Supply Paper 1819-B, 26 p.

Parks, W. S. (1973). *Geological map of the southwest Memphis quadrangle Tennessee*. U. S. Geological Survey Open-File Report, scale 1:24,000.

Parks, W. S. (1990). *Hydrogeology and preliminary assessment of the potential for contamination of the Memphis aquifer in the Memphis area, Tennessee*. U.S. Geological Survey Water-Resources Investigation Report 90-4092, 39 p.

Parks, W. S., Graham, D. D., Lowery, J. F. (1981). *Chemical character of ground water in the shallow water-table aquifer at selected localities in the Memphis area, Tennessee*. U. S. Geological Survey Open-File Report 81-223, 32 p.

Petracov, E. V. (1972). Geochemistry of fluoride in the groundwater of Moldavian Artesian Basin." *Ph.D Dissertation.* Leningrad State University, Russia, 280 p (in Russian).

Polev, P. V. & Negadaev-Nikonov, K. N. (1965). Geology of the USSR (Vol. 45, 455 p), Moldavskai SSR. Nedra, Moscow. (in Russian).

Precipitation. (2014). https://www.climate.gov/tags/precipitation.

Robinson, J. L., Carmichael, J. K., Halford, K. J., & Ladd, D. E. (1997). *Hydrogeologic framework and simulation of ground-water flow and travel time in the shallow aquifer system in the area of naval support activity Memphis, Millington, Tennessee*. U.S. Geological Survey Water-Resources Investigation Report 97-4228, 56 p.

Saraevskii, L. P. (1983). Groundwater resources of the territory between Rivers Prut and Nistru. *Ph.D Dissertation, All Soviet Union Institute of Hydrogeology and Engineering Geology*, 260 p (in Russian).

Saucier, R. T. (1994). *Geomorphology and quaternary geological history of the lower Mississippi valley* (365 p). U.S. Army Corps of Engineering.

Schneider, R. & Cushing, E. M. (1948). "Geology and Water Bearing Properties of the "1400-foot" Sand in the Memphis Area." In: *U.S. Geological Survey Circular* 33, 3 p**.

Sinthov, I. F. (1882). *Geological investigations of Basarabia and same parts of Herson region: Materials for geology of Russia* (Vol. 11, 12 p) (in Russian).

Soil Survey, Shelby County. (1989). Tennessee, United States Department of Agriculture, Soil Conservation Service in Cooperation with Tennessee Agricultural Experimental Station, 53 p.

Stasev, M. P. (1961). *Regional estimation of the fresh groundwater resources of Moldova*. AGeoM report, 150 p (in Russian).

Strucov, G. (1852). Groundwater of Basarabia: Journal of the state estates, No 3 (in Russian).

Tacis. (2000). *Prut water management* (1000 p). Report, Chisinau, IGS ASM.

TDEC (Tennessee Department of Environment and Conservation Division of Water Supply). (2002). Ground Water 305B Water Quality Report, Nov., 35p.

The Comprehensive Planning Section. (1981). Natural and Physical Characteristics of Memphis and Shelby County, Tennessee. The Shelby County Printing Department: 85 p.

USGS (United States Geological Survey). (2003). http://waterdata.usgs.gov/tn/nwis/gwlevels.

Van Arsdale, R. B., & Ten Brink, R. K. (2000). Late cretaceous and cenozoic geology of the New Madrid Seismic Zone. *Bulletin of the Seismological Society of America, 90*(2), 345–356.

Vzunzdaev, S. T. (1958). *Hydrogeohemical zonation of artesian water of the pre-Dobruja depression and slope of the Russian platform.* Report of the Academy of Sciences USSR, Vol. 114, No 4 (in Russian).

Vzunzdaev, S. T. (1965). *Hydrogeological investigations of the territory of Moldova.* Report IGG, Chisinau, 160 p (in Russian).

Waldron, B., Larsen, D., Hannigan R., & et al. (2011). Mississippi embayment regional ground water study. In: *EPA 600/R-10/130* (192 p).

Watzel, R., Ohnemus, J. (1997). Hydrogeologische Kartierung Karlsruhe – Speyer. Fortschreibung des Hydrogeologischen Baus im baden – wurttembergischen Teil. Gutachten des Geologischen Landesamtes Baden – Wurttemberg im Auftrag der Landesanstalt fur Umweltschutz Baden – Wurttemberg, AZ: 3531.01/96–4763.

Wolf, L., Eiswirth, M., Hotzl, H. (2003). Assessing sewer—ground water interaction at the city scale based on individual sewer defects (Vol. 50, No. 1, pp. 423–426). *RMZ—Materials and Geo-environment*, Ljubljana.

Wolf, L., Klinger, J., Hoetzl, H., & Mohrlok, U. (2007). Quantifying mass fluxes from urban drainage systems to the urban soil-aquifer system. *Journal of Soils and Sediments, 7*(2), 85–95.

Yearly trends: weather averages and extremes (Rastatt). (2014). http://www.myweather2.com/City-Town/Germany/Rastatt/climate-profile.aspx.

Zelenin, I. V. (1974). Investigation of filtration parameters of aquifers and legitimacy forming the Moldova groundwater resource. *Ph.D. Dissertation.* Moscow State University, 300 p. (in Russian).

Chapter 5
Hydrogeochemical Vulnerability Estimation

Constantin Moraru and Robyn Hannigan

Abstract Three representative test sites were used to determine the point of migration (PM) of pollutants in unsaturated zone. Test sites are located in different geographical, geological, and hydrogeological conditions. One common methodology was used for all study territories. In all cases, conclusions of the GAVEL methodology are confirmed with a preliminary study of groundwater quality. It is necessary to note that calculated PM is characteristic for a statistical period of time.

Keywords Test sites · Migration of pollutants · GAVEL methodology

5.1 Carpineni Test Site, Republic of Moldova

5.1.1 Descriptive Data

The test site named Carpineni is located in the southwest part of the Republic of Moldova in the frame of the Prut river basin (Fig. 5.1).

The size of the site is approximately 20×20 km. The density of the population is about 130 people/km^2. The entire population lives in rural conditions. Topographically, the investigated site is a relatively low-lying territory. The climate is moderate continental with a short mild winter and a long hot summer. In the cold season, there is a surplus of rainfall over evaporation as shown in Fig. 4.25. In the summertime, the situation is opposite with a deficiency of rainfall.

From a geological point of view, the Neogen and Quaternary strata build up the Carpineni site. They are exposed throughout the whole site, and only locally, some thin veneers of Quaternary alluvial sediments are found. Eolian loess deposits from the Pleistocene glacial periods are widespread, giving rise to the fertile black-colored loamy chernozem soils. In terms of Alpine Geology, this Middle to Late Miocene rocks (10–15 Mya) represent the so-called Molasse deposits of the Eastern Carpathian mountain ranges.

The Miocene strata and geological development have also dominated the geomorphological development during the Quaternary especially along the valleys.

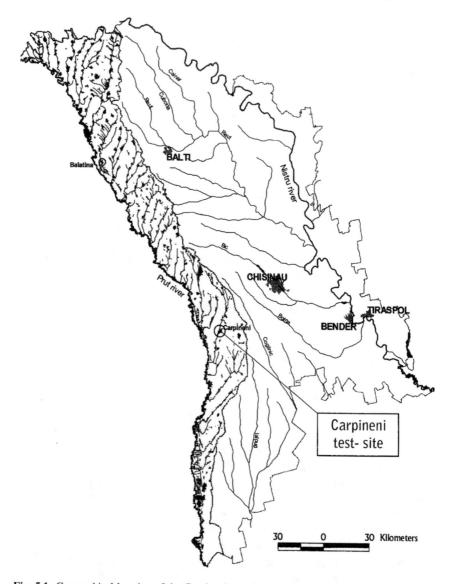

Fig. 5.1 Geographical location of the Carpineni test site

Spectacular mass movements and landslides are taking place continuously at a scale ranging from meters to several kilometers in these plastic clay Miocene sediments. Large-scale earth slides tend to push up a series of peculiar elongated hills along the front of the slide, giving rise to closed depressions filled with lakes and wetlands (Tacis 2000).

From a hydrogeological point of view, the overall predominance of clay sediments is very important with subordinate discontinuous limestone horizons (5–

15 m) and sand layers. The sediments almost always contain calcium carbonate; however, macrofossils are rare. Another important aspect is the fact that these marine sediments, in particular the clay strata, contain entrapped seawater from the time of deposition which, due to the geological history and development of a drainage pattern, has not been fully flushed.

Geological and hydrogeological structures of the investigated site are presented in Figs. 5.2, 5.3, and 5.4. Land use and the location of the agricultural sources of pollutions are shown in Figs. 5.5 and 5.6. Arable territories are used extensively for different agricultural purposes. Hydro-melioration is applied.

From well to well, the geochemistry of groundwater can vary significantly, and for an unconfined aquifer, water quality is correlated with the position of the water table. Statistical hydrogeochemical data are shown in Table 5.1.

Chemical composition of the groundwater is strongly dependent on TDS values. It was assumed that the dividing TDS value is 1.38 g/l. If TDS < 1.38 g/l, the water type is HCO_3–Na, and if TDS > 1.38 g/l, the water type is SO_4–HCO_3–Mg–Ca (by Kurlov classification). Generally, water polluted locally by NO_3 has high values of hardness and high values of some trace elements such as Se (Fig. 5.7). Configuration of the Se concentration is typical for all pollutants, which are located in and around villages. This fact is connected with the style of living in villages: no pipelines, anthropogenic remains are collected locally, etc. Dependence of water quality parameters and land use is presented in Fig. 5.8. From the box plots, it can be seen that the main sources of pollution are urban (i.e., here rural) territories and urban–arable areas.

5.1.2 GAVEL Application

The locations of the investigated wells are shown in Fig. 5.6. A line of four wells was drilled in the unsaturated zone, and samples were collected. The wells are located in different land use and geological–hydrogeological conditions. Lithology and natural variation of the TDS are not the same for all wells (Fig. 5.9). The unsaturated zone is built up by sand, clay, and sandy clay. From a hydrodynamic point of view, the water permeability of the rocks is not high. The TDS varies mainly depending on the type of rocks. Nevertheless, TDS variation is statistically difficult to understand and to determine the type of function. In Fig. 5.9, the averaged line is a six-parameter polynomial function without a clear statistical tendency.

The initial geochemical data for the unsaturated zone are summarized in Table 5.2. Cumulative transformation of data is presented in Table 5.3. Cumulative graphs and PM determination are shown in Figs. 5.10, 5.11, 5.12, and 5.13. The PM or l_k can be determined only for Wells 1, 2, and 3 (Table 5.4). This is because Well 4 penetrated the shallow aquifer at the small depth.

Data from Table 5.4 are used for the calculation of the GAVEL parameter R (Table 5.5). According to the data from this table, risk of groundwater pollution is

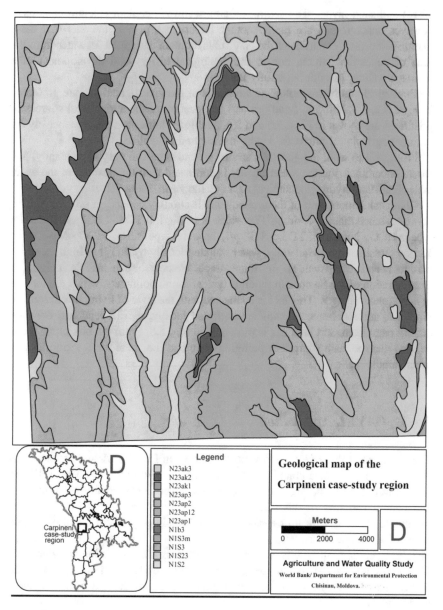

Fig. 5.2 Geological map of the Carpineni case study region

low. Values of PM are in the interval from 2.27 to 2.30, and values of R are between 0.13 and 0.19. Taking into consideration that Wells 1, 2, and 3 are located on the watershed, and on medium part of the slope, using a method of geological analogy, the following can be concluded. All territories on the case study site

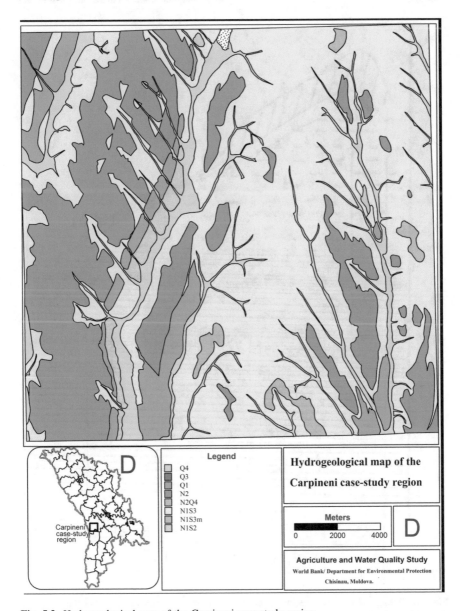

Fig. 5.3 Hydrogeological map of the Carpineni case study region

topographically selected by the line of the medium slope are at the low risk of vulnerability. Areas located below this line can be considered at risk for pollution. An illustrative argument is shown in Fig. 5.7. Selenium pollution is strongly located only in the low topographical relief. Here, the thinness of the unsaturated zone is not more than 2.0–3.0 m. Pollution of shallow groundwater with nitrogen

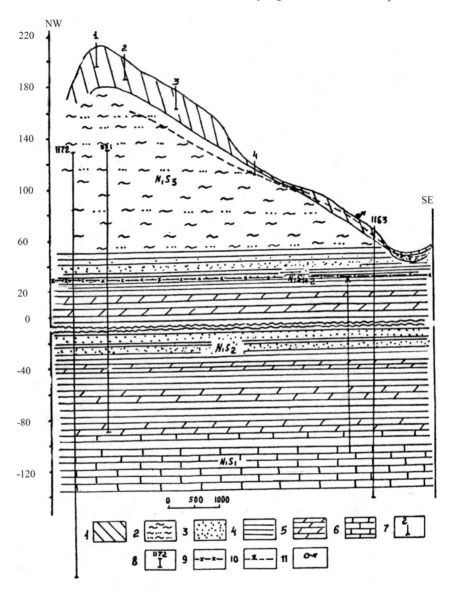

Fig. 5.4 Geological and hydrogeological cross section of the Carpineni test site (*1* Sandy clay, *2* Sandwich of sand and clay, *3* Sand, *4* Clay, *5* Marl, *6* Limestone, *7* Investigation well, *8* Water supply well, *9* Water table of confined aquifers, *10* Water table of unconfined aquifer, and *11* Spring)

substances (like NO_2 and NO_3) has the same distribution. TDS values are distributed in the same manner as well. In these areas, a dense network of dug wells is used for freshwater supply. Water quality in the most cases does not correspond to hygienic standards and needs special treatment.

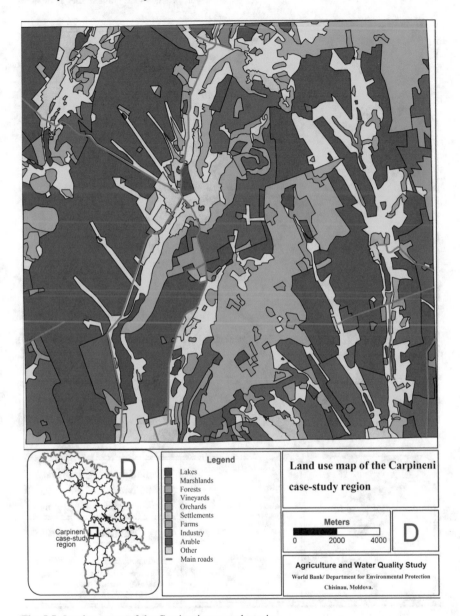

Fig. 5.5 Land use map of the Carpineni case study region

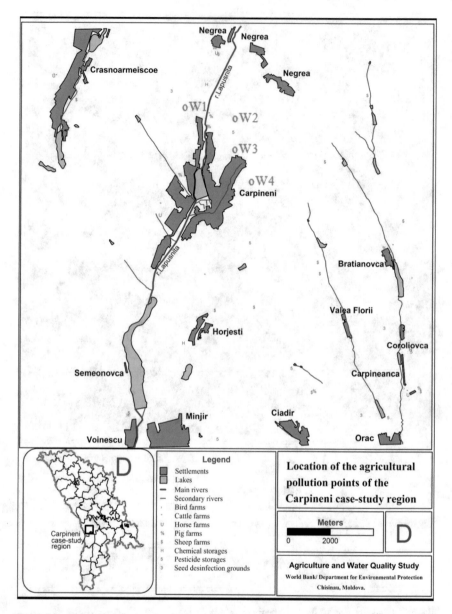

Fig. 5.6 Location of agricultural pollution points in the Carpineni case study region (red W1–W4 are wells drilled for investigation)

Table 5.1 Statistical data for groundwater of the unconfined aquifer, Carpineni site

Parameter	Number of samples	Mean	Standard deviation	Minimum	Maximum
K, mg/l	19	4.47	6.06	0.6	22.40
Na, mg/l	19	141.01	67.97	9.00	240.00
Ca, mg/l	19	69.80	67.84	5.00	243.00
Mg, mg/l	19	62.63	41.60	1.26	179.80
HCO$_3$, mg/l	19	476.76	179.24	85.40	793.00
SO$_4$, mg/l	19	204.98	270.42	2.13	1000.80
NO$_3$, mg/l	19	41.18	33.37	2.67	106.70
Cl, mg/l	19	71.66	62.78	13.62	211.00
Oxidizability, mg/l	19	2.23	2.82	0.40	13.10
pH, unit	19	7.73	0.61	6.70	9.45
TDS, mg/l	19	840.70	429.60	258.00	2000.00

5.2   Rastatt Test Side, Germany

5.2.1   Descriptive Data

The test site is located in the town of Rastatt, Germany (Fig. 5.14). The population reaches approximately 50,000 with about 35,000 residents living in the urban center of the administrative area. The urban area of Rastatt covers about 15 km^2. Climatic and geographical conditions were described in Chap. 4. It is important to note that groundwater in this area is used for freshwater supply. The main source of pollution is the sewer system. The entire length of the sewer system in Rastatt reaches approximately 200 km and consists of separated (30%) and combined sewers (70%) (Eiswirth et al. 2003). The city is built on Quaternary gravel and sand sediments. A detailed geological cross section is shown in Fig. 5.15. Usually, shallow groundwater is located in the frame of Pleistocene terrace gravel.

For the city of Rastatt, the overall simulated contaminant fluxes from the undisturbed daily operation of the urban drainage system are too small to cause a widespread groundwater contamination, at least for the parameters of boron and chloride (Wolf et al. 2007).

5.2.2   GAVEL Application

Three wells were drilled in the frame of the Rastatt area (for location, see Fig. 5.14). Samples were collected from different intervals of the unsaturated zone. In the laboratory, water extractions were performed. For each sample, electrical conductivity (EC) was measured. Values for EC were recalculated in the TDS

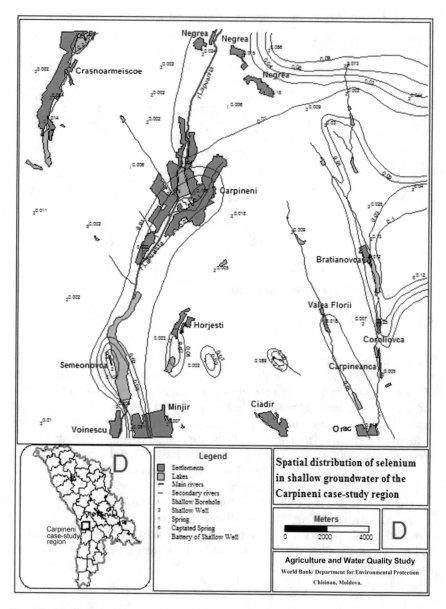

Fig. 5.7 Distribution of Se (mg/l) in unconfined aquifer, Carpineni site

concentrations. For this purpose, calibration solutions were prepared, and the TDS equation was obtained (Fig. 5.16). Such transformations are widely used in geochemistry, for example, see Fig. 5.17.

Initial data for EC and calculated TDS are shown in Table 5.6. From the beginning, it necessary to note that TDS values in the unsaturated zone are highly

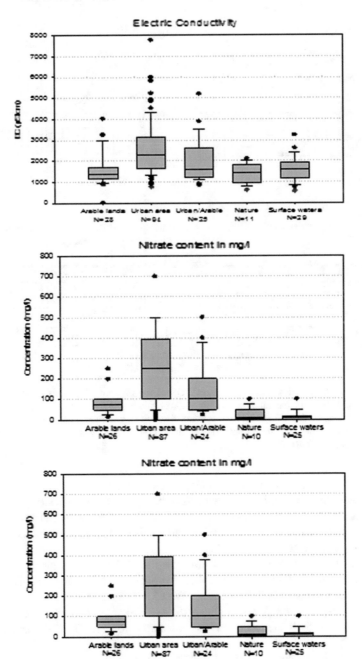

Fig. 5.8 Boxplots of electrical conductivity, nitrate, and chloride in shallow groundwater for different land use types (*N* is the number of samples used for geostatistics)

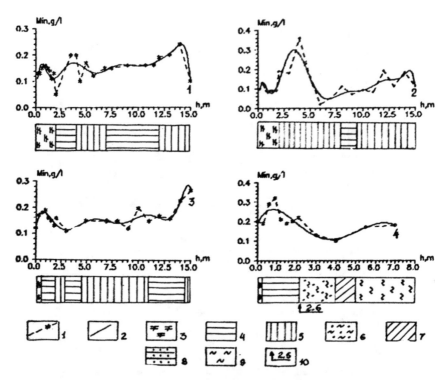

Fig. 5.9 Lithology and natural variation of TDS of the unsaturated zone, Carpineni test site, Moldova (*1* Line of the natural TDS variation, *2* Averaged function of TDS, *3* Soil, *4* Yellow lam, *5* Red dark lam, *6* Clay with small sand stratum, *7* Sandy clay, *8* Lam with sand stratum, *9* Clay, and *10* Water table and its value; number near TDS line is well number, Min is water mineralization or TDS)

variable. The main factors determining such variations are stratigraphy of the unsaturated zone in combination with its hydrodynamic properties (Fig. 5.18).

Cumulative transformations of TDS values are summarized in Table 5.7. Graphs of the natural variation of TDS, and its cumulative representation including FPM determination, are shown in Fig. 5.19.

Analysis of data from Fig. 5.19 indicates that PM was determined exactly for only Wells 1 and 2. The depth of the Well 3 is not enough, because technically deep drilling was not possible. In the case of Well 3, PM is considered to be more than approximately 1.80 m. Accordingly to the GAVEL methodology (see Chap. 4), data from Fig. 5.19 are used for the calculation of the GAVEL parameter R (Table 5.8).

Water level of the unconfined aquifer varies in the small interval, namely between 0.5 and 2.0 m. In the bottom of unsaturated zone, compressed sand and

Table 5.2 Geochemical data for Carpineni test site (h, m; chemical elements—mg/l)

Well	h	pH	HCO$_3$	Cl	SO$_4$	K	Na	Ca	Mg	NO$_3$	TDS
W1	0.2	42773	122	24624	42786	42888	7	43099	30348	n.d.	198.62
	0.4	42773	97.6	42807	42784	42737	42861	34.68	21245	n.d.	183.16
	0.6	42954	122	42807	42908	42796	6	39.94	42828	n.d.	217.34
	0.8	42832	146.4	42807	42847	42796	4	43.44	30376	n.d.	243.07
	1	42832	122	42807	42791	42888	42950	46.24	12145	n.d.	224.87
	42767	42186	122	42807	42787	42888	4	39.58	42859	n.d.	214.33
	42826	12966	122	42807	42788	42768	42950	36.44	24198	n.d.	211.95
	42887	42862	122	24624	42959	42826	42949	42758	27820	n.d.	179.58
	42948	42742	122	42807	42956	42796	42858	36.44	21276	n.d.	199.02
	2	7	97.6	42807	42804	42826	3	43092	21276	n.d.	160.3
	42858	42801	122	42783	42916	42826	42992	13.75	42756	n.d.	228.16
	4	34881	122	42783	42916	42826	42992	13.75	42756	n.d.	228.81
	42829	42743	122	42984	42773	42736	6	10	17.42	n.d.	178.72
	5	45839	109.8	42783	26	42767	42960	42871	15.18	n.d.	206.93
	42921	42801	109.8	42870	42971	42887	42930	7	19.24	n.d.	199.94
	42953	42773	103.7	42960	42851	42917	42840	41487	42784	n.d.	194.35
	42862	42923	103.7	42960	42913	16438	17	42801	42933	n.d.	195.62
	9	42801	122	42960	42970	42795	42791	42862	15.18	n.d.	216.08
	42926	16619	128.1	42870	42913	1	40	45839	28004	n.d.	236.66
	42866	45108	140.3	42894	42895	12785	39.9	45870	45809	n.d.	221.48
	12	42954	122	42906	42915	27395	46.4	42864	23894	n.d.	244.3
	13	42801	134.2	42960	42846	42826	54.2	7	11810	n.d.	244.62
	14	23924	158.6	42960	42910	42795	70.6	5	36220	n.d.	285.54
	15	42801	158.6	42804	10	42795	56	27607	36220	n.d.	256.24
W2	0.2	42954	73.2	13.34	42781	42826	4	28.38	30713	n.d.	146.16
	0.4	45839	97.6	26.66	42899	42887	42952	34.3	15401	n.d.	190.23
	0.8	42862	79.3	44835	7	15342	20515	26.28	42768	n.d.	137.03
	1	7	73.2	33909	42802	21186	42920	21.78	27030	n.d.	130.22
	42767	42801	97.6	24990	42985	42768	21947	22.86	21245	n.d.	150.72
	42826	16619	97.6	13.34	42953	42767	42827	20	30742	n.d.	152.63
	42887	42186	97.6	24990	42772	42767	2	42813	42859	n.d.	144.63
	42948	7	97.6	24990	42772	42767	42827	20	42860	n.d.	146.58
	2	42773	91.5	42996	32.4	42767	42836	25.75	25051	n.d.	197.03
	3	16619	103.7	42870	28	1	42836	23.25	14185	n.d.	201.68
	4	42773	91.5	42870	31.2	42767	42958	17.25	30987	n.d.	187.49
	5	42832	134.2	42984	42959	42767	42953	16.13	42994	n.d.	201.52
	6	42985	79.3	42984	7	42917	42745	42798	42744	n.d.	126.3
	7	42186	97.6	42984	42834	42767	42833	41579	42747	n.d.	153.88
	8	16619	109.8	42984	42834	42005	42833	32325	42869	n.d.	165.48
	9	22828	85.4	42984	42833	42826	42984	14062	28399	n.d.	134.77
	10	23924	146.4	42771	42952	42979	42984	41609	18.89	n.d.	204.87
	11	42743	146.4	42771	42952	42948	42954	45901	18.89	n.d.	203.24
	42866	43077	85.4	42984	42773	42917	10	6	30987	n.d.	137.16
	13	42801	103.7	42984	42773	42856	10	41426	13.82	n.d.	156.55
	14	15523	134.2	42984	42830	2	42804	25416	17.29	n.d.	192.2
	15	7	85.4	42984	42895	42826	42894	27607	45566	n.d.	137.89

(continued)

Table 5.2 (continued)

Well	h	pH	HCO₃	Cl	SO₄	K	Na	Ca	Mg	NO₃	TDS
W3	0	42953	85.4	16954	42954	42889	42831	42822	42769	30682	149.8
	0.3	34851	85.4	25447	12	42768	42773	33.5	42798	0.6	161.84
	0.6	31959	122	44986	42754	0.6	3	39.8	42890	11324	201.51
	0.9	42802	128.1	16954	42932	1	42828	44.2	42771	42769	216.56
	42767	42863	122	16954	42742	0.8	3	36.4	42860	42156	195.46
	42856	42801	115.9	16954	23437	0.7	42828	34.6	42922	26543	188.42
	42948	42773	115.9	25447	34759	1	42828	32.6	42743	42948	182.92
	2	45870	128	25447	42891	1	42828	42914	42867	0.4	197.44
	3	27576	122.1	16954	42859	42826	42889	42932	42812	42918	182.88
	4	12966	128.2	16954	42743	42856	42771	42962	42908	42736	196.31
	42982	12997	134.2	16954	42866	42767	42892	42958	42759	0.79	205
	42984	42743	128.1	16954	42807	42826	42862	42958	42880	0.4	202.56
	7.00	42773	128.1	27576	20	42736	42864	42928	42880	0.4	212.25
	8	42773	128.1	16954	5	1	42783	13	42750	0.97	194.03
	9	42832	128.1	16954	42796	0.7	42886	42986	42862	32509	194.75
	10	41487	183	44986	42797	0.9	47	11	42743	42827	266.7
	11	45108	122	44986	42833	42767	38	42894	42919	44927	193.59
	12	42801	85.4	16954	42872	42795	42885	42984	42769	0.5	159.06
	13	42833	134.2	16954	42828	1	45.7	42864	42858	0.4	212.56
	14	12997	122	16954	42807	0.9	41.8	42833	42796	23012	205.14
	15	42217	128.1	16954	17	0.9	45.7	42833	42796	42737	219.11
W4	0	42893	158.6	3	42830	42923	2	43.36	34029	42952	237.39
	0.3	16619	164.7	3	42891	42887	42955	39	22706	42831	240.17
	0.6	20271	158.6	29.2	23	42826	42746	50.39	8	42783	306.44
	0.9	22828	146.4	42972	42773	42856	42953	43.86	17.37	65	321.55
	42767	8	146.4	12	42891	42736	42866	25.38	16.23	42939	249.91
	42856	22828	164.7	42894	42774	42736	42752	42815	16.61	42782	261.16
	42948	31959	170.8	42828	42894	42826	32.5	41883	14.95	42980	251.56
	42737	27576	176.9	42771	42894	42736	32.5	42804	14.95	42983	263.2

Remark n.d. is no data

gravel are deposited. Water permeability of these deposits is low. The risk of pollution from land surface, in terms of CAVEL classification, is medium to low. Confirmation of this fact can be depicted from maps of electrical conductivity and chloride (Figs. 5.20 and 5.21).

Simple comparison of the tested well location (Fig. 5.14) and Figs. 5.20 and 5.21 show the following. For medium and low vulnerability, EC values range from 300 to 500 uS/cm and chloride concentrations 15.0–20.0 mg/l. Eiswirth et al. (2003) noted that the distribution of the specific electric conductivity (SEC) of the Rastatt groundwater in the probability net is normal, and the medians of the conductivity of the three data records are in between 544 and 575 µS/cm. The

Table 5.3 Cumulative geochemical data for Carpineni test site (h, m; chemical elements—mg/l)

Well	h	Cl	NO₃	HCO₃	SO₄	K	Na	Ca	Mg	TDS
W1	0.2	24624	n.d.	122.00	20.20	21947	7.00	43099	30348	191.42
	0.4	19.97	n.d.	219.60	38.40	25659	13.50	64.80	15128	367.38
	0.6	33.27	n.d.	341.60	61.00	7.00	19.50	104.74	29830	576.92
	0.8	46.57	n.d.	488.00	83.40	11202	23.50	148.18	13.64	812.59
	1	59.87	n.d.	610.00	108.60	33178	27.30	192.42	17.97	1028.06
	42767	73.17	n.d.	732.00	129.80	14.50	31.30	232.00	22.47	1235.24
	42826	86.47	n.d.	854.00	152.00	16.70	35.10	268.44	27.13	1439.84
	42887	93.14	n.d.	976.00	164.80	43026	37.90	291.54	30.89	1612.37
	42948	106.44	n.d.	1098.00	174.60	20.30	41.40	327.98	35.47	1804.19
	2	119.74	n.d.	1195.60	184.90	21.70	44.40	351.10	40.04	1957.48
	42858	136.94	n.d.	1317.60	215.50	43031	59.30	364.85	61.06	2178.35
	4	154.14	n.d.	1439.60	246.10	24.50	74.20	378.60	82.07	2399.21
	42829	161.04	n.d.	1561.60	253.30	25.60	80.20	388.60	99.49	2569.83
	5	178.24	n.d.	1671.40	279.30	26.80	94.00	405.10	114.67	2769.51
	42921	193.74	n.d.	1781.20	304.10	28.40	108.70	412.10	133.91	2962.15
	42953	207.54	n.d.	1884.90	330.50	43038	124.10	420.23	151.93	3149.30
	42862	221.34	n.d.	1988.60	358.10	31.55	141.10	427.53	169.00	3337.22
	9	235.14	n.d.	2110.60	381.90	32.85	166.30	435.03	184.18	3546.00
	42926	250.64	n.d.	2238.70	409.50	33.85	206.30	442.28	193.94	3775.21
	42866	259.24	n.d.	2379.00	419.10	35.20	246.20	450.53	200.19	3989.46
	12	279.84	n.d.	2501.00	448.70	36.95	292.60	460.03	206.84	4225.96
	13	293.63	n.d.	2635.20	470.10	38.35	346.80	467.03	212.16	4463.27
	14	307.44	n.d.	2793.80	494.70	39.65	417.40	472.03	216.15	4741.17
	15	317.74	n.d.	2952.40	504.70	40.95	473.40	480.78	220.14	4990.11
W2	0.2	13.34	n.d.	73.20	15.20	14611	4.00	28.58	30713	138.56
	0.4	40	n.d.	170.80	28.80	14702	29465	62.88	46174	322.94
	0.8	50.22	n.d.	250.10	35.80	30072	13.36	89.16	17015	452.92
	1	62.14	n.d.	323.30	44.10	14793	42904	110.94	44105	576.14
	42767	68.82	n.d.	420.90	52.00	22160	20.66	133.80	13.78	719.56
	42826	82.16	n.d.	518.50	58.80	29495	42909	153.80	17.62	864.74
	42887	88.84	n.d.	616.10	65.00	12.00	42911	173.10	43091	1002.22
	42948	95.52	n.d.	713.70	71.20	13.20	27.46	193.10	27.62	1141.80
	2	114.42	n.d.	805.20	103.60	14.40	38.86	218.85	36.30	1331.63
	3	129.92	n.d.	908.90	131.60	15.40	50.26	242.10	47.68	1525.86
	4	145.42	n.d.	1000.40	162.80	16.60	62.06	259.35	59.52	1706.15
	5	152.32	n.d.	1134.60	175.60	17.80	68.86	275.48	75.61	1900.27
	6	159.22	n.d.	1213.90	182.60	19.50	78.96	279.78	85.32	2019.28
	7	166.12	n.d.	1311.50	192.00	20.70	87.36	290.91	97.42	2166.01
	8	173.02	n.d.	1421.30	201.40	21.85	95.76	298.79	111.92	2324.04
	9	179.92	n.d.	1506.70	209.80	23.25	102.66	306.17	122.69	2451.19
	10	185.12	n.d.	1653.10	215.60	25.15	109.56	318.30	141.58	2648.41
	11	190.32	n.d.	1799.50	221.40	26.95	117.36	327.55	160.47	2843.55
	42866	197.22	n.d.	1884.90	228.60	28.65	127.36	333.55	172.31	2972.59
	13	204.12	n.d.	1988.60	235.80	30.15	147.36	339.68	186.13	3131.84
	14	42787	n.d.	2122.80	241.20	32.15	157.66	348.37	203.42	3126.62
	15	217.92	n.d.	2208.20	250.80	33.55	166.26	357.12	213.66	3447.51

(continued)

Table 5.3 (continued)

Well	h	Cl	NO_3	HCO_3	SO_4	K	Na	Ca	Mg	TDS
W3	0	16954	30682	85.40	29403	21976	14763	28.30	43891	143.00
	0.3	16.15	16103	170.80	19.80	29342	13.60	61.80	18445	297.89
	0.6	19.38	27454	292.80	38.90	14763	16.60	101.50	43020	491.43
	0.9	25.84	34851	420.90	55.60	14793	20.00	145.80	17.30	699.79
	42767	32.3	43021	542.90	62.70	44044	23.00	182.20	22.80	887.20
	42856	38.76	22.82	658.80	66.34	33086	26.40	216.80	29.50	1068.32
	42948	48.45	23.90	774.70	70.29	33117	29.80	249.40	37.60	1244.04
	2	58.14	24.30	902.70	79.89	33147	33.20	278.00	50.10	1437.23
	3	64.6	26.37	1024.80	80.39	11293	36.80	294.70	68.40	1608.36
	4	71.06	27.47	1153.00	88.49	13.80	42.00	310.50	91.00	1797.32
	42982	77.52	28.26	1287.20	99.99	15.00	48.60	322.30	115.10	1993.97
	42984	83.98	28.66	1415.30	113.29	16.40	56.10	334.10	140.60	2188.43
	7	91.73	42915	1543.40	133.29	17.50	17.00	346.80	166.10	2344.88
	8	98.19	42824	1671.50	138.29	18.50	34.20	359.80	181.20	2531.71
	9	104.65	31.92	1799.60	140.59	19.50	65.70	368.70	188.70	2719.36
	10	107.88	33.96	1982.60	143.89	43028	112.70	379.70	196.80	2977.63
	11	111.11	35.19	2104.60	152.29	21.30	150.70	388.30	200.50	3163.99
	12	117.57	35.69	2190.00	169.79	22.60	181.20	395.20	203.70	3315.75
	13	124.03	36.09	2324.20	173.19	23.60	226.90	404.70	207.20	3519.91
	14	130.49	37.72	2446.20	186.49	24.50	268.70	413.10	209.50	3716.70
	15	136.95	39.82	2574.30	203.49	25.40	314.40	421.50	211.80	3927.66
W4	0	3	29342	158.60	14732	25750	2.00	43.36	34029	229.79
	0.3	6	44166	323.30	11.00	11202	29495	82.36	20271	462.51
	0.6	35.2	29.40	481.90	34.00	25842	21.90	132.75	15.55	761.40
	0.9	61	94.40	628.30	41.20	44166	28.70	176.61	32.92	1075.33
	42767	73	118.10	774.70	46.80	13.30	40.20	201.99	49.15	1317.24
	42856	81.6	134.30	939.40	55.00	14.40	57.30	223.02	65.76	1570.78
	42948	85	137.20	1281.00	63.60	15.80	89.80	232.16	80.71	1985.27
	42737	90.2	143.10	1457.90	72.20	16.90	122.30	242.46	95.66	2240.72

Remark n.d. no data

beginning of the unambiguously anthropogenically influenced area starts with 760 µS/cm in the Rastatt model area. Maximum specific conductivity values have been recorded up to 1941 µS/cm. The same authors continue that the distribution of the chloride concentration in Rastatt groundwater is crooked in the probability net, and the chloride medians of the three present data records are between 18 and 20.5 mg/l. The beginning of the unambiguously anthropogenically influenced area starts with 46 mg/l chloride in the Rastatt model area. Maximum chloride concentrations have been recorded up to 1566 mg/l.

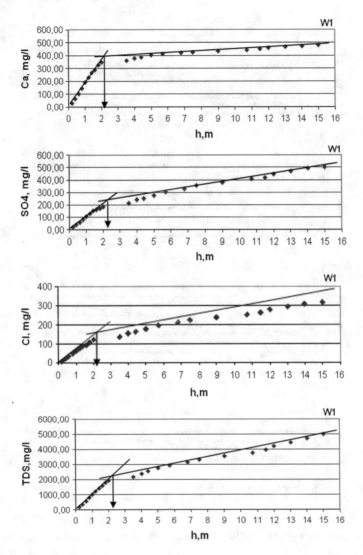

Fig. 5.10 Cumulative graphics and FPM determination for Well 1 (arrow indicates the value of the PM = l_k)

5.3 Memphis Area, USA

5.3.1 Descriptive Data

The Memphis area is a part of Shelby County, Tennessee. The investigated area, from a natural point of view, is complex because of different soil and geological

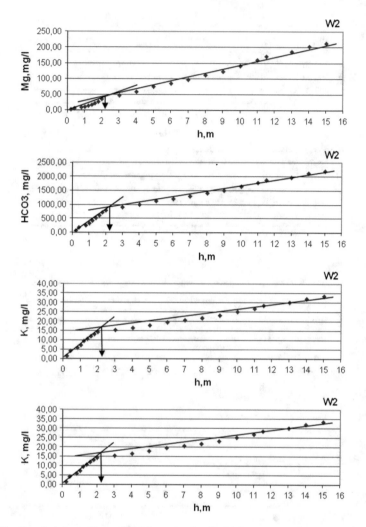

Fig. 5.11 Cumulative graphics and PM determination for Well 2 (arrow indicates the value of the PM = l_k)

conditions, land use practices, and hydrogeological structure. The north part of the Shelby County was selected for detailed study (Fig. 5.22).

Shallow groundwater is more vulnerable to surface pollution. Water table aquifers in the Memphis area consist of the alluvium and fluvial deposits, which are mostly unconfined (Brahana and Broshears 2001). A map of generalized surface geology (Fig. 5.23) presents the distribution of alluvium and fluvial deposits (as loess) (Chang and Hwang 1989).

Alluvium occurs in the stream valleys. The alluvium is not a major groundwater source in the Memphis area, even though it is a major water-bearing zone and can

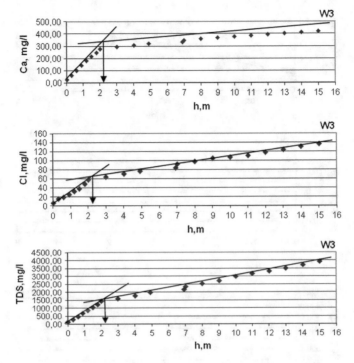

Fig. 5.12 Cumulative graphs and PM determination for Well 3 (arrow indicates the value of the PM = l_k)

supply large quantities of water to wells. West, north, and south of the study area, the alluvium of the Mississippi River alluvial plain is one of the most productive regional aquifers in the Mississippi embayment, supplying over a billion gallons per day to irrigation wells in Arkansas and Mississippi (Boswell et al. 1968). In the Mississippi River alluvial plain, the alluvium is commonly 100–175 feet thick (Boswell et al. 1968). Along valleys of upland streams and tributaries to the Mississippi River east of the bluffs, thickness generally is less than 50 feet (Brahana and Broshears 2001). Lithologically, alluvium includes gravel, sand, silt, and clay; the latter is commonly rich in organic matter.

Fluvial deposits occur at land surface in the uplands. Fluvial deposits range in thickness from 0.0 to 100.0 feet. Thickness is highly variable because of surfaces at both the top and base (Brahana and Broshears 2001). The lithology of fluvial deposits is primarily sand and gravel, with minor layers of ferruginous sandstone. Fluvial deposits are separated from the Memphis aquifer by sediments of the Jackson Formation and the upper part of the Claiborne group.

Both alluvium and fluvial deposits contain groundwater. One unique Quaternary unconfined aquifer is characteristic for these stratigraphic units. Water from this aquifer is used mostly for agriculture. The Quaternary aquifer is hydraulically

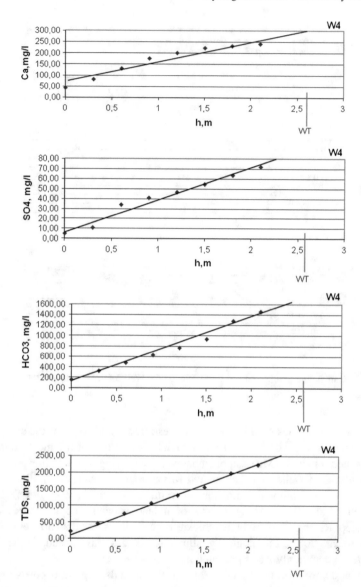

Fig. 5.13 Cumulative graphs for Well 4 (vertical red line indicates the value of aquifer water table WT)

connected with the Memphis sand one, which is regionally used for freshwater supply. This fact presents an ecological interest.

Geochemistry of the unsaturated zone of the Memphis area, as a whole, and particularly at the test site is insufficient. Limited data exist in USGS and EPA archives, as well in official published materials. We selected the northern part of the

Table 5.4 Determination of the PM (in m) for tested wells, Carpineni site, Moldova

Parameter	PM for well 1	PM for well 2	PM for well 3
Ca	2.20	–	–
SO_4	2.30	–	–
Cl	2.30	–	–
TDS	2.30	–	–
Mean	2.27	–	–
Mg	–	2.30	–
HCO_3	–	2.30	–
K̄	–	2.30	–
TDS	–	2.30	–
Mean	–	2.30	–
Cl	–	–	2.30
Ca	–	–	2.20
Cl	–	–	2.30
TDS	–	–	2.30
Mean	–	–	2.27

Table 5.5 Calculation of the risk of groundwater pollution and GAVEL classification

Well	Water table (m)	l_k (m)	R	CAVEL classification
1	17	2.27	0.13	Risk is low
2	16	2.30	0.14	Risk is low
3	12	2.27	0.19	Risk is low
4	2.6	2.6	1.00	Polluted water

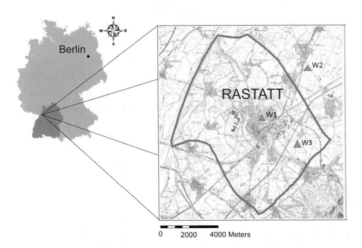

Fig. 5.14 Location of the Rastatt test site (Wolf et al. 2007 and modified by authors) (blue triangles W1–W3 are wells drilled for investigation)

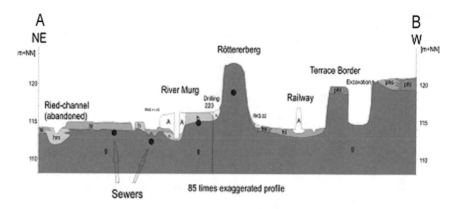

Fig. 5.15 Cross section in the urban area of Rastatt, redrawn after (Wolf et al. 2007). g = Pleistocene terrace gravel; phs = Pleistocene floodplain sand; phl = Pleistocene floodplain loam; hl = Holocene loam; h = Holocene undifferentiated; hs = Holocene sand; hm = Holocene meander fill; and A = anthropogenic fill

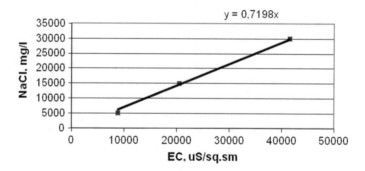

Fig. 5.16 Calibration line of the TDS (as NaCl) and EC

Shelby County, which is located near Millington town. The selected site is well-documented from a pedology point of view. The investigated test site is represented at the small-scale in Fig. 5.24. In our case, the best way to represent the unsaturated zone is detailed soil map (Web soil survey 2014). Typical soil profiles are shown in Fig. 5.25.

5.3.2 GAVEL Application

Thickness of the unsaturated zone, in the frame of site territory, is 3.0 m on average. Information about mineralogical and chemical composition is scarce and only for the upper part of the zone. Chemical soil properties were used for computing and the assessment of PM (l_k). Selected data for chemical soil properties are

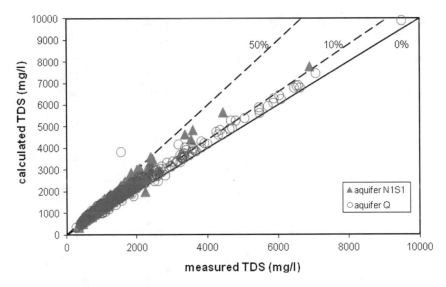

Fig. 5.17 Correlation between measured and calculated TDS, Moldova (Jousma et al. 2000)

summarized in Table 5.9 (Web soil survey 2014). Authors modified this table. Depth means the point of sampling in the soil profile. Cation-exchange capacity is the total amount of extractable cations that can be held by the soil, expressed in terms of milliequivalents per 100 grams of soil at neutrality (pH = 7). Effective cation-exchange capacity refers to the sum of extractable cations plus aluminum expressed in terms of milliequivalents per 100 grams of soil. It is determined for soils that have a pH of less than 5.5. The ability to retain cations reduces hazard of groundwater pollution. pH is the soil reaction to acidity or alkalinity (Web soil survey 2014).

In Table 5.10, data (from Table 5.9) were selected and used for GAVEL method. Only effective cation-exchange capacity (ECEC) is representative and statistically suitable for the GAVEL method. On the basis of these data, cumulative graphs were constructed for the ECEC including a graph for pH (Figs. 5.26, 5.27, 5.28). Analysis of data from Tables 5.9 and 5.10, and Figs. 5.26, 5.27, and 5.28 suggest the following information. Cumulative distribution of ECEC is not composed of two functions. Only one trend line is characteristic. This means that migration of cations and aluminum is more than depths of 1.5 m (GaB soil), 2.0 m (He soil), and 1.4 m (GaC3 soil). In all cases, changes of pH take place at this depth. Geochemically, such changes are connected with the natural acid-alkaline state of the unsaturated zone. Based on obtained information, we can assume that the probable value of PM is, on average, between 2.0 and 2.5 m (2.25 m). Calculation of the vulnerability of groundwater (in this case unconfined) is equal to:

Table 5.6 Initial data for the Rastatt site, Germany

Well	Depth (m)	EC (uS/sq.cm)	TDS (mg/l)
1	0.00	58.00	41.74
	0.10	70.00	50.38
	0.30	87.00	62.62
	0.50	93.00	66.94
	0.70	82.00	59.02
	0.90	92.00	66.22
	1.10	75.00	53.98
	1.30	91.00	65.50
	1.60	56.00	40.30
	2.15	67.00	48.22
	2.45	57.00	41.02
	3.50	70.00	50.38
	4.25	77.00	55.42
	4.50	68.00	48.94
	5.00	46.00	33.11
2	0.00	71.00	51.10
	0.10	52.00	37.42
	0.30	25.00	17.99
	0.50	19.00	13.67
	0.70	27.00	19.43
	0.90	29.00	20.87
	1.15	12.00	8.63
	1.50	28.00	20.15
	1.70	9.00	6.47
	1.90	10.00	7.20
	2.90	18.00	12.95
	3.25	13.00	9.35
	3.85	12.00	8.63
	4.50	25.00	17.99
3	0.00	32.00	23.03
	0.10	35.00	25.19
	0.30	17.00	12.23
	0.50	15.00	10.79
	1.10	15.00	10.79
	1.30	12.00	8.63
	1.50	18.00	12.95
	1.80	17.00	12.23
	2.25	12.00	8.63

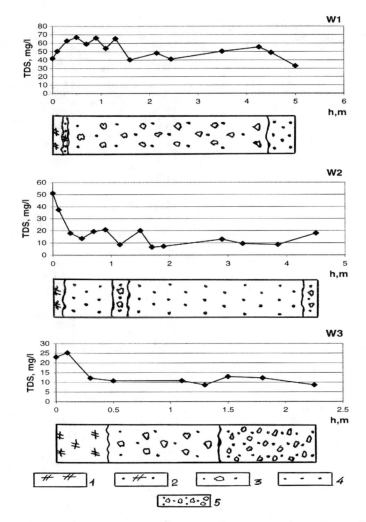

Fig. 5.18 Lithology and natural variation of the TDS of the unsaturated zone, Rastatt, Germany (*1* Soil, *2* Soil and sand, *3* Sand with gravel, *4* Sand, and *5* Compressed sand and gravel)

$$R = l_k/H_s = 2.25/3 = 0.75$$

In terms of groundwater vulnerability, conditions where R is 0.5 < R < 0.75 will be at high risk of groundwater pollution (HR). This assumption (because a lack of geochemical data) is proven by the hydrogeochemical conditions of the investigated territory. Here, high values of TDS (more than 150 mg/l) are characteristic. For surrounding territories, TDS is less than 100 mg/l. High values of TDS are associated with anomalies of Cl, Na, Mg, Zn, F, and Si (Moraru and Anderson 2005). Such hydrogeochemical conditions are formed only under the influence of

Table 5.7 Cumulative TDS data, Rastatt, Germany

Well	Depth (m)	TDS (mg/l)
1	0.00	41.74
	0.10	92.12
	0.30	154.74
	0.50	221.68
	0.70	280.70
	0.90	346.92
	1.10	400.90
	1.30	466.40
	1.60	506.70
	2.15	554.92
	2.45	595.94
	3.50	646.32
	4.25	701.74
	4.50	750.68
	5.00	783.79
2	0.00	51.10
	0.10	88.52
	0.30	106.51
	0.50	120.18
	0.70	139.61
	0.90	160.48
	1.15	169.11
	1.50	189.26
	1.70	195.73
	1.90	202.93
	2.90	215.88
	3.25	225.23
	3.85	233.86
	4.50	251.85
3	0.00	23.03
	0.10	48.22
	0.30	60.45
	0.50	71.24
	1.10	82.03
	1.30	90.66
	1.50	103.61
	1.80	115.84
	2.25	124.47

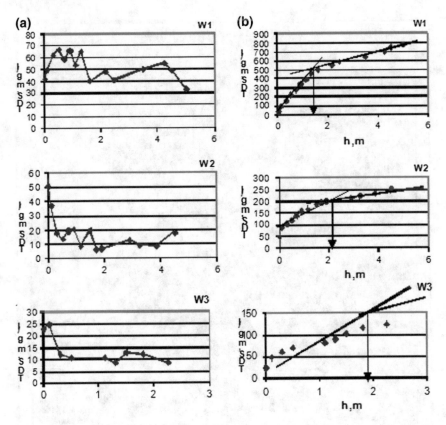

Fig. 5.19 Natural variations of the TDS (**a**) and (**b**) TDS cumulative distribution including PM determination (arrow indicates the value of the PM = l_k)

Table 5.8 Calculation of the risk of groundwater pollution and GAVEL classification

Well	Water table (m)	l_k (m)	R	CAVEL classification
1	5.31	1.4	0.26	Risk is medium
2	8.45	1.8	0.21	Risk is low
3	7.50	1.8	0.24	Risk is low

anthropogenic pollution. Main sources of pollution are at the top of unsaturated zone. Transportation of pollutants is done by water from atmospheric precipitations. Similar conditions of hydrogeochemical vulnerability, in a general view, are characteristic of all of the Memphis area.

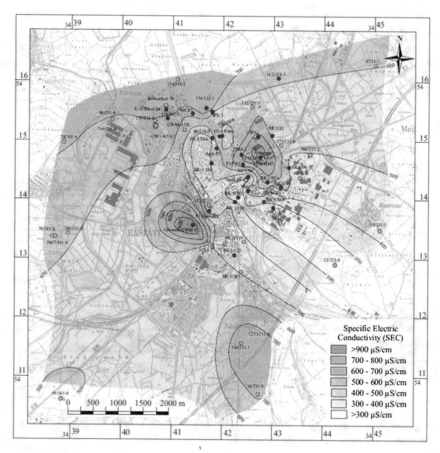

Fig. 5.20 Unconfined groundwater electrical conductivity of the Rastatt area (Eiswirth et al. 2003)

5.4 Discussion

Theoretical background of the PM is well-described in Chap. 2. Three representative test sites were used to determine the point of migration of pollutants in unsaturated zone. Test sites are located in different geographical, geological, and hydrogeological conditions. One common methodology was used for all study territories.

For the Carpineni test site (Republic of Moldova), geochemistry of the unsaturated zone is well-studied. The studied territory is a site of agricultural land use practices. Data are available and analytically are accurate. From the large spectrum of chemical elements determined in the soil and rocks of unsaturated zone, only major elements, TDS, NO_3, and pH, were selected. These elements are characteristic for different intervals of the zone. Finally, employing of the GAVEL

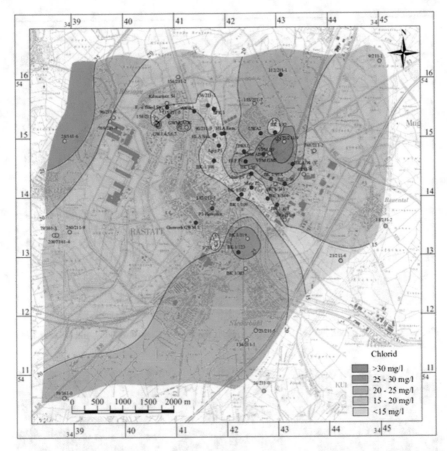

Fig. 5.21 Unconfined groundwater chloride distribution of the Rastatt area (Eiswirth et al. 2003)

Fig. 5.22 Location of the
Memphis area (rectangle is
the position of test site)

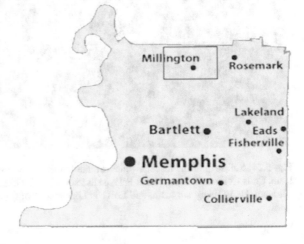

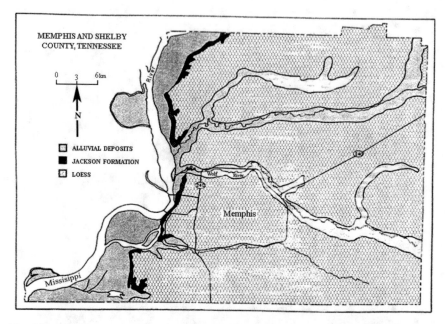

Fig. 5.23 Map of generated subsurface geology of the Shelby County (Chang and Hwang 1989)

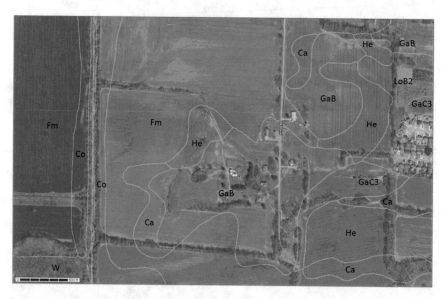

Fig. 5.24 Soil map of the test site (lines are the borders between soil types; Ca is Calloway silt loam, Co is Collins silt loam, Fm is Falaya silt loam, GaB is Grenada silt loam, GaB3 is Grenada silt loam, He is Henry silt loam, and LoB2 is Loring silt loam) (Web soil survey 2014)

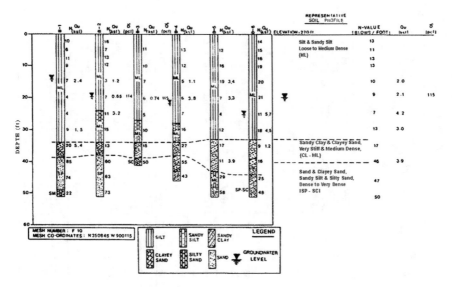

Fig. 5.25 Representative soil profiles (Chang and Hwang 1989)

methodology permits the determination of very accurate values of the PM. According to GAVEL classification, the vulnerability of groundwater (first of all unconfined aquifer) is low, and in one well, a value of $R = 1.0$ means that the aquifer is already polluted.

Rastatt test site (Germany) is completely different compared to the Carpineni site. The main difference is in the type of land use, urban versus town, and sources of pollution. Special wells were drilled, and from core material, samples were prepared. The decision was made to use the integral parameter TDS as an indicator of unsaturated zone pollution. TDS has been computed using measuring electrical conductivity of water solutions for each sampled interval. Statistically, the data were enough for the planned task. Analytical accuracy is high. Calculation of the vulnerability risk by GAVEL methodology indicates that risk of pollution is from low to medium.

The Memphis area (USA) is well-characterized geochemically. The test site belongs to agricultural and urban conditions. Soil cover is described pedologically and by main chemical characteristics. Effective cation exchange capacity (ECEC) (sum of extractable cations and aluminum) was used for our purposes. Statistical data for unconfined water table, and generalized data for ECEC, allow for the conclusion that vulnerability risk of groundwater pollution is high ($R < 0.75$).

In all cases, conclusions of the GAVEL methodology are confirmed by a preliminary study of groundwater quality. It is necessary to note that calculated PM is characteristic for a statistical period of time. Usually, this time consists several decades. In the case of changing of the potential properties of the water sources at the top of unsaturated zone, PM will vary in the frame of the unsaturated zone. For long periods of time (20,000–120,000 years), the unsaturated zone serves as an

Table 5.9 Chemical soil properties

Map symbol and soil name	Depth (in.)	Average depth (in.)	Average depth (m)	Cation-exchange capacity (meq/100 g)	Effective cation-exchange capacity (ECEC) (meq/100 g)	Average ECEC (meq/100 g)	Soil reaction (pH)	Average pH
Ca—Calloway silt loam	0–7	42858	27607		40300	45809	4.5–6.0	45778
Calloway	44378	14	35.00		42372	42864	4.5–6.0	45778
	21–60	40.5	101.25	42743		42867	5.1–7.8	16589
Co—Collins silt loam	0–8	4	10		2.0–7.3	23833	4.5–5.5	5.00
Collins	18476	28	70		0.9–7.2	12875	4.5–5.5	5.00
Fm—Falaya silt loam	0–60	30	75		1.9–9.4	42125	4.5–5.5	5.00
Falaya	0–7	42858	27607		2.5–9.4	34790	4.5–5.5	5.00
Waverly	22098	33.5	83.75		0.8–6.1	16497	4.5–5.5	5.00
GaB—Grenada silt loam	0-6.0	3	42862		3.1–5.0	42859	4.5–6.0	45778
Grenada	44713	14	35		40638	42923	4.5–6.0	45778
	22–26	24	60		3.8–7.3	20210	4.5–6.0	45778
	26–40	33	82.5		42617	16711	4.5–3.0	27454
	40–60	50	125	42893		42806	5.1–7.3	43983
GaC3—Grenada silt loam	0–6	3	42862		3.1–5.1	42739	4.5–6.0	45778
Grenada	44713	14	35		40638	42923	4.5–6.1	11079
	22–26	25	62.5		3.8–7.3	20210	4.5–6.0	45778
	26–40	33	82.5		42617	16711	4.5–6.0	45778
	40–60	50	125	42893		42806	5.1–7.3	43983

(continued)

Table 5.9 (continued)

Map symbol and soil name	Depth (in.)	Average depth (in.)	Average depth (m)	Cation-exchange capacity (meq/100 g)	Effective cation-exchange capacity (ECEC) (meq/100 g)	Average ECEC (meq/100 g)	Soil reaction (pH)	Average pH
He—Henry silt loam	0–9	42859	45962		2.2–5.7	31107	4.5–5.5	5.00
Henry	44075	42869	36.25		3.4–6.3	31138	4.5–5.5	5.00
	20–60	40	100		41522	16681	4.5–5.5	5.00
	60–90	75	187.5	40578		42893	5.1–7.8	16589
LoB2—Loring silt loam	0.0–7	42858	27607				4.5–6.0	45778
Loring	46935	42872	43.75				4.5–6.0	45778
	28–50	36	90				4.5–6.8	42125
	50–60	55	137.5				4.5–6.5	18384

Table 5.10 Data for chemical soil properties using the GAVEL method

Map symbol and soil name	Average depth (m)	Average ECEC (meq/100 g)	Cumulative ECEC (meq/100 g)	Average pH
GaB—Grenada silt loam Grenada	7.5	4.05	4.05	5.25
Grenada	35	7.7	11.75	5.25
	60	5.55	17.30	5.25
	82.5	10.45	17.75	3.75
	125	12.3	40.05	6.20
GaC3—Grenada silt loam	7.5	4.1	4.10	5.25
Grenada	35	7.7	11.80	5.30
	62.5	5.55	17.35	5.25
	82.5	10.45	27.70	5.25
	125	12.3	40.00	6.20
He—Henry silt loam	11.25	3.85	3.85	5.00
Henry	36.25	4.85	8.70	5.00
	100	9.45	18.15	5.00
	187.5	7.6	25.75	6.45

Fig. 5.26 Cumulative distribution of ECEC (**a**) and normal values of pH (**b**) for Grenada soil type

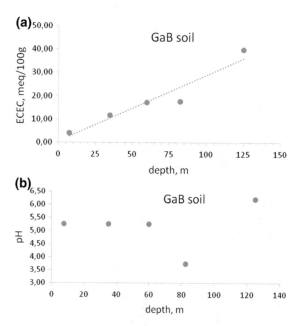

Fig. 5.27 Cumulative
distribution of ECEC (**a**) and
normal values of pH (**b**) for
Henry soil type

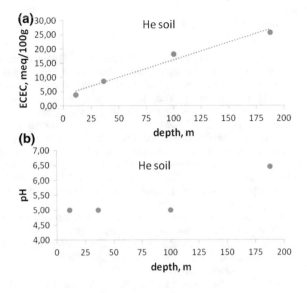

Fig. 5.28 Cumulative
distribution of ECEC (**a**) and
normal values of pH (**b**) for
Grenada soil type

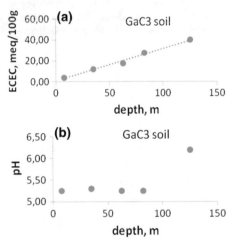

archive of past climate. Edmunts and Tyler (2002) compared the Cl profiles with instrumental records, such as rainfall and rivers, gauging records, and ^{3}H profiles. Model studies have helped to define the persistence time of unsaturated zone signals, where evidence of a 20-year event, such the Sanel drought, may persist for 1000 years.

References

Boswell, E. H., Cushing, E. M., & Hosman, R. L. (1968). Quaternary aquifers in the Mississippi Embayment. US Geological Survey Professional Paper 448-E, p. 15.

Brahana, J. V., & Broshears, R. E. (2001). Hydrogeology and ground-water flow in the Memphis and Fort Pillow aquifers in the Memphis area, Tennessee. US Geological Survey Water-Resources Investigation Report 89-4131, p. 56.

Chang, K. W. Hg., & Hwang, H. H. M. (1989). Subsurface conditions of Memphis and Shelby County. Technical Report NCEER 89-0021 (July 26 1989), p. 30.

Edmunts, W. M., & Tyler, S. W. (2002). Unsaturated zones as archive of past climates: Towards a new proxy for continental regions. *Hydrogeology Journal, 10,* 216–228.

Eiswirth, M., Held, I. I., Wolf, L., & Hotzl, H. (2003). AISUWRS work package 1. Background study. University of Karlsruhe, Department of Applied Geology, Commissioned report 30.08.2003, p. 78.

Jousma, G., Kloosterman, F., Moraru, C., et al. (2000). Groundwater and land use. Report of the TACIS Prut water management project, p. 180.

Moraru, C. E., & Anderson, J. A. (2005). Comparative assessment of the ground water quality of the Republic of Moldova and the Memphis, TN area of the United States of America. Ground Water Institute, Memphis, TN, p. 188.

Tacis. (2000). Prut water management. Report, Chisinau, IGS ASM p. 1000.

Web Soil Survey. (2014). http://websoilsurvey.nrcs.usda.gov/app/HomePage.htm.

Wolf, L., Klinger, J., Hoetzl, H., & Mohrlok, U. (2007). Quantifying Mass Fluxes from urban drainage systems to the urban soil-aquifer system. *Journal of Soils and Sediments, 7*(2), 85–95.

Chapter 6
Groundwater Geochemistry and Vulnerability

Constantin Moraru and Robyn Hannigan

Abstract On the basis of compositional data theory, a new hydrogeochemical classification is proposed. This classification is used for the comparison of the groundwater quality assessment for the Republic of Moldova, the Memphis, Tennessee, USA area and Rastatt, Germany. An assessment of groundwater quality for both the Memphis area and the Republic of Moldova, as well as the Rastatt area in Germany, according to the international standards and the Moraru classification is presented.

Keywords Hydrogeochemical classification · Groundwater quality Groundwater vulnerability

6.1 Moraru Hydrogeochemical Classification

The authors are introducing a new classification for groundwater. Integrated analyses show that different classifications are in use throughout the world. A new hydrogeochemical classification is proposed here for the comparison of the groundwater quality assessment for the Republic of Moldova, the Memphis, Tennessee, USA area and Rastatt, Germany. The proposed techniques describe better and more logically, how groundwater is vulnerable to pollution (Moraru and Anderson 2005).

6.1.1 Theoretical Background

It is well known that only six major constituents dominate the chemical composition of the hydrosphere, and groundwater in particular, namely the anions Cl, SO_4, HCO_3 and the cations Na, Ca, Mg (Posohov 1975; Samarina 1977; Pitiova 1978, 1984; Hem 1985). These constituents form 95 to 98% of the TDS, and they essentially contribute to the geochemical properties of water itself. At low-temperature conditions, geochemical associations of the major elements in

© Springer International Publishing AG 2018
C. Moraru and R. Hannigan, *Analysis of Hydrogeochemical Vulnerability*,
Springer Hydrogeology, https://doi.org/10.1007/978-3-319-70960-4_6

conjunction with pH-Eh conditions determine the forms in which the remaining chemical ingredients, such as trace elements, can occur in the water.

Many researchers mention the relationship between TDS and major chemical constituents in aqueous media (Valiashko 1939; Chebotarev 1955; Posohov 1975; Samarina 1977). The functional relationship of TDS to anion and cation contents is well known. Samarina (1977) and Pitiova (1978) used the geochemical dependence of major elements from TDS in their proposed classifications. However, neither these nor other classifications simultaneously show the direct influence of TDS to water type.

Aitchison (1981, 1986) introduced the theory and practice of compositional data. Consequently, for this publication, TDS is treated as a compositional parameter. The main assumption of this theory is that any vector $\mathbf{x}$ with nonnegative elements x_1 .... x_D that represents proportions of some whole is subject to the obvious constraint:

$$X_1 + \cdots + X_D = 1 \quad \text{or, equivalently, to } 100\%$$

TDS is a compositional parameter consisting essentially (95–98%) of Cl, SO_4, and HCO_3, and Na, Ca, and Mg. The remaining 2–5% is attributable to inorganic and organic constituents, and their significance in TDS is negligible. The data can be arranged in a format as illustrated in Table 6.1.

The matrix of data from Table 6.1 has the following properties:

1. Each row of data corresponds to a single sample.
2. Each column of data corresponds to only one chemical constituent.
3. The sum of $Mn_1 + \cdots + Mn_6 = TDSn_i = 1$ (or = 100%), i.e., the sum of entries in each row is 1 (or 100%) and represents the conventional value of TDS.

Such properties are characteristic features of a compositional data set (Aitchison 1986). Figure 6.1 shows the graphical representation of the compositional sample 1 from Table 3.12.

In fact, Fig. 6.1 is a linear vector with length equal to TDS. Following the principle of predominant anion and predominant cation, nine unique water types can be obtained as illustrated in Table 6.2.

6.1.2 Hydrochemical Parameters and Classification Scheme

The Moraru classification uses six major chemical elements, including HCO_3, SO_4, Cl (anions) and Na, Ca, Mg (cations), as well as TDS. The Moraru classification initially expresses parameters in mg/l. After analytical determination of major chemical compounds in a water sample, the classification scheme requires the following steps:

Table 6.1 Hydrogeochemical compositional data

Sample	Chemical constituents						TDS
	Na	Ca	Mg	HCO$_3$	SO$_4$	Cl	
1	M1	M2	M3	M4	M5	M6	TDS1
...	...	...	...	...	...	...	...
m	Mn$_1$	Mn$_2$	Mn$_3$	Mn$_4$	Mn$_5$	Mn$_6$	TDSn$_i$

where 1, ..., and m are the number of samples ($m > 0$); M1...M6 and Mn$_1$...Mn$_6$ are concentrations of chemical constituents as parts of TDS, $0.0 <$ values < 1.0 or $0.0 <$ values $< 100.0\%$; TDS1... TDSn$_i$ are total dissolved solids, with statistical values $= 1.0$ or 100.0%

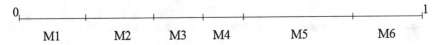

Fig. 6.1 Graphical view of a compositional data

Table 6.2 Groundwater types of the Moraru classification

Anions	Cations		
	Na	Ca	Mg
HCO$_3$	NaHCO$_3$	CaHCO$_3$	MgHCO$_3$
SO$_4$	NaSO$_4$	CaSO$_4$	MgSO$_4$
Cl	NaCl	CaCl	MgCl

1. Select HCO$_3$, SO$_4$, Cl and Na, Ca, Mg from a chemical analysis protocol. Values are in mg/l.
2. Select TDS in mg/l from the same protocol.
3. Calculate a new TDS as: TDS $= (HCO_3/2) + SO_4 + Cl + Na + Ca + Mg$.
   Note: HCO$_3$ is divided by two because bicarbonate is converted to carbonate in the solid phase. The bicarbonate is generally converted by a gravimetric factor as mg/l HCO$_3 \times 0.4917 =$ mg/l CO$_3$ (Hem 1985).
4. Compare the new TDS and the TDS from the protocol. The new TDS values may differ from TDS, which is a residue on evaporation. If this difference is equal to $\pm 20\%$, for the TDS of the order <3000 mg/l, go to the next step. If the difference is more than 20%, the chemical analysis is not accurate and not usable for classification.
5. Calculate the percentage of HCO$_3$, SO$_4$, Cl, and Na, Ca, Mg from the newly calculated TDS, using the assumption that TDS is equal to 100%.
6. Select a predominant anion and a predominant cation, and write the water type according to Table 6.2.

Table 6.3 Qualitative groups of groundwater according to total dissolved solids values (Chebotarev 1955 and modification by the authors)

Major groups	Water index	Subdivisions	TDS (g/l)
Freshwater	F1	Good potable	<0.5
	F2	Fresh	0.5–0.7
	F3	Fairly fresh	0.7–1.5
	F4	Passably fresh	1.5–2.5
Brackish water	B1	Slightly brackish	2.5–3.2
	B2	Brackish	3.2–4.0
	B3	Definitely brackish	4.0–5.0
Saltwater	S1	Slightly salt	5.0–6.5
	S2	Salt	6.5–7.0
	S3	Very salt	7.0–10.0
	S4	Extremely salt	>10.0

Remark Water index is spelling *F1* fresh first category, *B1* brackish first category, etc

7. Select the water index, as presented in Table 6.3.
8. Write the Moraru formula of the water classification in the format:

Water Index–Water Type

For example: F2 $CaHCO_3$, B1 $MgSO_4$, etc. and the corresponding phrase: fresh second category calcium bicarbonate water; brackish first category magnesium sulfate water.

Classifications can be used to map various water types in cross sections for practical and scientific purposes. Figure 6.2 is the proposed graphical presentation of the Moraru classification. The nomograph of the Moraru classification consists of a nonscaled vertical axis of TDS, which is divided into the three major classes—fresh, brackish, and saline—with divisions into subclasses. The chemical compositional data of TDS are plotted on six vectors and add up to 100%. The TDS value is plotted in mg/l, and then the percentage of HCO_3, SO_4, Cl and Na, Ca, Mg is added. Finally, all points are connected in one polygonal line. From the nomograph, it is possible to read water quality and TDS class (value). For example: in Fig. 6.2, sample 1 belongs to F1 $NaHCO_3$; sample 2—F3 $CaSO_4$; and sample 3—B3 NaCl.

6.2   Groundwater Geochemistry and Vulnerability

Following is an assessment of groundwater quality for both the Memphis area and the Republic of Moldova, as well as the Rastatt area in Germany, according to the standards and the Moraru classification (proposed in this publication). In 2007, the Republic of Moldova approved a drinking water standard (DWS), which still prevails in the country. This DWS has three parts: microbiological, chemical, and

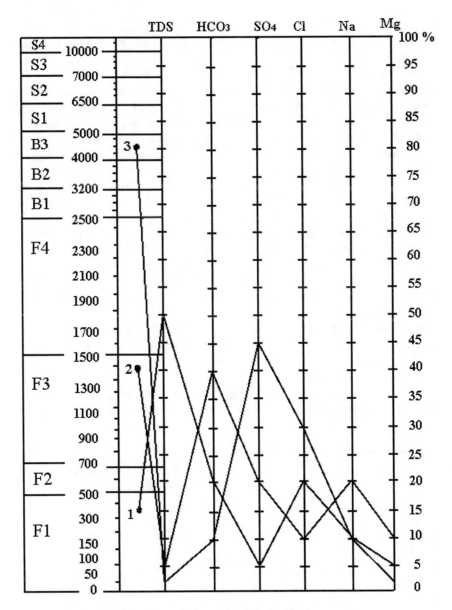

Fig. 6.2 Nomograph of the Moraru hydrogeochemical classifications

organoleptic indicators. In Tennessee, in 2003, the primary and secondary EPA (Environmental Protection Agency) water standards were applied. The primary standards are enforced and must be met by all public drinking water providers. Secondary standards are not enforced, but are maintained to protect the public welfare and to ensure a supply of pure, wholesome, and potable water. In Germany, the Drinking Water Directive (DWD) of EU (1998) applies. The DWD is also composed of three parts, namely microbiological, chemical, and indicator parameters.

The World Health Organization (WHO) has published the guidelines for drinking water quality (WHO 2011). This is the most internationally recognized document for assessing water quality and contains detailed recommendations and guidelines for water quality. The guidelines are composed of an introduction, guideline requirements, health-based targets, a water safety plan, surveillance, other applications of the guidelines, microbial aspects, chemical aspects, radiological aspects, and acceptability aspects.

There are 10 representative standard constituents used to assess groundwater quality, according to the DWS, EPA, WHO, and DWD EU: pH, total dissolved solids (TDS), total iron (Fe), manganese (Mn), arsenic (As), fluoride (F), chloride (Cl), nitrite (NO_3), sulfate (SO_4), and selenium (Se) and are presented in Table 6.4.

Data from Table 6.4 indicate that only values for arsenic (As), chloride (Cl), and nitrate (NO_3) are the same for all four standards. Differences in values for the remaining elements are more significant. The EPA standard, considering both MCL or TT and secondary recommendation, integrates values that are lower compared to DWS, WHO, and EU standards.

Statistical water quality assessments for the Memphis area, the Republic of Moldova, and Rastatt, Germany, are each, respectively, illustrated in Tables 6.5, 6.6, and 6.7. Figures 6.3, 6.4, and 6.5 show nomographs according to the Moraru classification.

A detailed analysis of groundwater quality assessment using the EPA, DWS, EU, and WHO standards suggests the following. The shallow groundwater of the

Table 6.4 Representative standard values of the DWS, EPA, WHO, and EU

No	Parameter	Unit	EPA, USA		DWS, Moldova	WHO, International	European Union, including Germany
			MCL or TT	Secondary			
1	pH	units		6.5–8.5	6.5–9.5	D 6.5–8.5	6.5–9.5
2	TDS	mg/l		500	1500	D 1000	D 1000
3	Fe	mg/l		0.3	0.3	D 0.3	0.2
4	Mn	mg/l		0.05	50	D 0.5	0.05
5	As	mg/l	0.01		0.01	0.01	0.01
6	F	mg/l	4	2	42,856	42,856	42,856
7	Cl	mg/l		250	250	D 250	250
8	NO_3	mg/l	50		50	50	50
9	SO_4	mg/l		250	250	D 250	250
10	Se	mg/l	0.05		0.01	0.04	0.01

Remark MCL is maximum contamination level, and TT is treatment technique intended to reduce the level of the contaminant; blank spaces are without parameters values determined by EPA, DWS, and WHO; D is desirable concentration. WHO outline for these parameters: not at health concern at levels found in drinking water

Table 6.5 Statistical matrix of groundwater quality assessment, Memphis area, USA (mg/l)

Shallow aquifer parameter	N	Minimum	Maximum	EPA min	EPA max	DWS min	DWS max	WHO min	WHO max	EU min	EU max
SO₄	85	0.00	200.00	n.e.	n.e	n.e	n.e	n.e	n.e	n.e.	n.e.
Cl	85	0.00	280.00	n.e.	+1.12	n.e.	+1.72	n.e.	+1.12	n.e.	+1.12
TDS	85	41.00	1180.00	n.e.	+2.36	n.e.	n.e.	n.e.	+1.18	n.e.	+1.18
F	85	0.00	0.70	n.e.	n.e.	n.e.	n.e.	n.e.	n.e.	n.e.	n.e.
Fe	54	0.00	80.00	n.e.	+267	n.e.	+267	n.e.	+267	n.e.	+400
Mn	85	0.00	14.00	n.e.	+280	n.e.	n.e.	n.e.	+140	n.e.	+280
pH	82	5.10	7.60	−1.27	n.e.	−1.17	n.e.	n.e.	−1.17	−1.17	n.e.

Remark n.e. is no excess; minus (−) indicates that parameter value is lower compared with the standard; plus (+) indicates that parameter value is greater than the standard; As, NO₃, and Se are not shown in the table because of absence of data; *N* is the amount of samples

Table 6.6 Statistical matrix of water quality assessment, Republic of Moldova (mg/l)

Shallow aquifer	N	Minimum	Maximum	EPA min	EPA max	DWS min	DWS max	WHO min	WHO max	EU min	EU max
NO₃	405	0.00	1908.00	n.e	+38.18	n.e.	+42.42	n.e.	+42.4	n.e.	+42.4
SO₄	428	2.13	4899.84	n.e.	+19.6	n.e.	+19.6	n.e.	+19.6	n.e.	+19.6
Cl	428	3.00	1539.40	n.e.	+6.15	n.e.	+6.15	n.e.	+6.15	n.e.	+6.15
As	10	0.002	0.015	n.e.	+1.5	n.e	+1.5	n.e.	+1.5	n.e.	+1.5
F	268	0.000	10.500	n.e.	+2.6	n.e	+7	n.e.	+7	n.e.	+7
Fe	121	0.011	0.300	n.e.	+1	n.e.	+1	n.e.	+1	n.e.	+1.5
Mn	117	0.001	0.740	n.e.	+14.8	n.e.	n.e.	n.e.	+7.4	n.e.	+14.8
Se	102	0.002	0.250	n.e.	+5	n.e.	+25	n.e.	+6.25	n.e.	+25
TDS	428	252.80	9478.59	n.e.	+18.95	n.e.	+9.5	n.e.	+6.3	n.e.	+9.5
pH	413	6.500	10.240	n.e.	+1.2	n.e.	+1.13	n.e.	+1.07	n.e.	+1.07

Remark see—Table 6.5

Table 6.7 Statistical matrix of water quality assessment, Rastatt, Germany (mg/l)

Shallow aquifer	N	Minimum	Maximum	EPA min	EPA max	DWS min	DWS max	WHO min	WHO max	EU min	EU max
NO_3	118	0.4	50.7	n.e.	+1.01	n.e.	+1.01	n.e.	+1.01	n.e.	+1.01
SO_4	118	19.9	73.0	n.e.	n.e.	n.e.	n.e.	n.e.	n.e.	n.e.	n.e.
Cl	118	8.7	52.0	n.e.	n.e.	n.e	n.e.	n.e.	n.e	n.e	n.e
As	118	0.001	0.005	n.e.	n.e.	n.e.	n.e.	n.e.	n.e.	n.e.	n.e.
F	118	0.02	0.45	n.e.	n.e.	n.e	n.e.	n.e.	n.e.	n.e.	n.e.
Fe	118	0.01	18.6	n.e.	+62	n.e.	+62	n.e.	+62	n.e.	+93
Mn	118	0.001	2.7	n.e.	+54	n.e.	n.e.	n.e.	+27	n.e.	+54
Se	118	n.d.	n.d.	n.d.	n.d.	n.d.	n.d.	n.d.	n.d.	n.d.	n.d.
TDS	118	96.36	515.0	n.e.	n.e.	n.e.	n.e.	n.e.	n.e.	n.e.	n.e.
pH	118	6.16	7.8	n.e.	n.e.	n.e.	n.e.	n.e.	n.e.	n.e.	n.e.

Remark see—Table 6.5; *n.d.*—absence of data

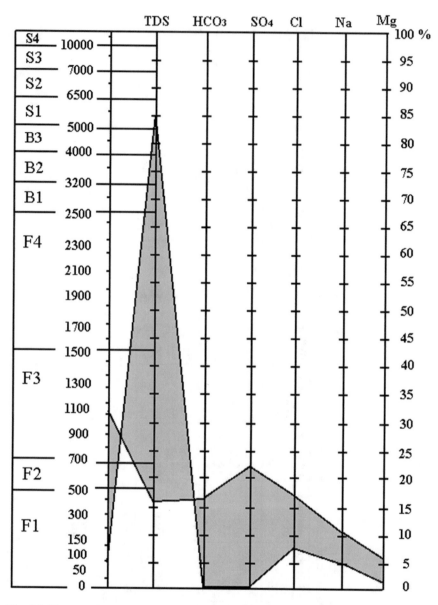

Fig. 6.3 Nomograph of the Moraru classification shallow aquifer, Moldova

Memphis area contains excessive Cl, Fe, Mn, and TDS and low pH values. For Moldova's groundwater, the spectrum of contaminants is more significant. The shallow aquifer contains elevated values of all selected parameters. The unconfined aquifer from Rastatt, Germany contains elevated concentrations of nitrate, iron, and manganese.

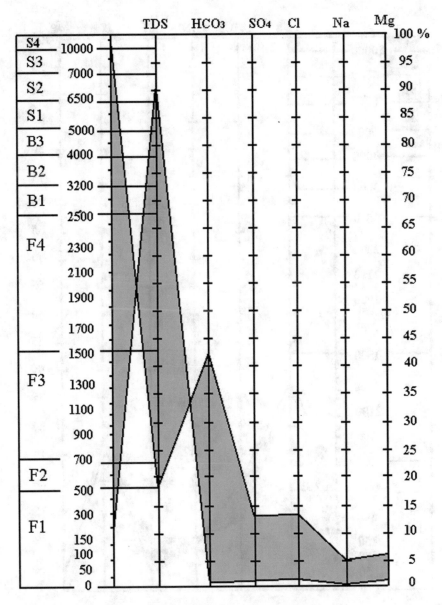

Fig. 6.4 Nomograph of the Moraru classification, shallow aquifer, Memphis area

TDS is the general indicator of the geochemical state of groundwater. Anomalies are visible for Moldova, the Memphis area, and Rastatt territory in the maps of regional TDS distribution (Figs. 6.6, 6.7, and 6.8; 5.20).

For the Memphis area, such anomalies have mostly a mosaic character (Fig. 6.6). In the north part of Shelby County, TDS is regularly distributed and its

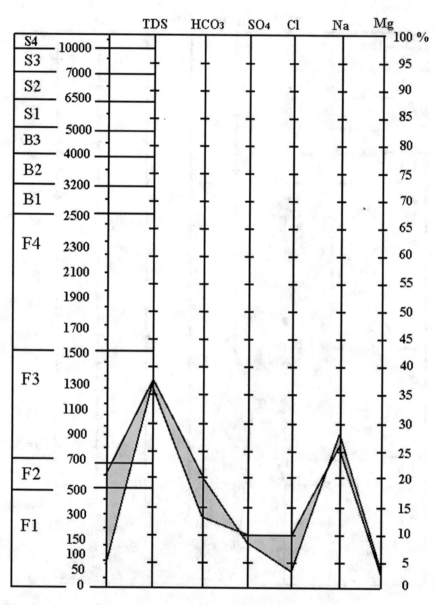

Fig. 6.5 Nomograph of the Moraru classification, shallow aquifer, Germany

value changes from 100 to 250 mg/l. This territory was investigated in detail, and groundwater vulnerability is determined to be at high risk ($0.5 < R < 0.75$). The distribution of calcium (Ca) for the same area (Fig. 6.7) indicates that the vulnerable territories have a high concentration of Ca (between 15 and 50 mg/l). Such calcium concentrations in the unconfined aquifer are only due to infiltration of

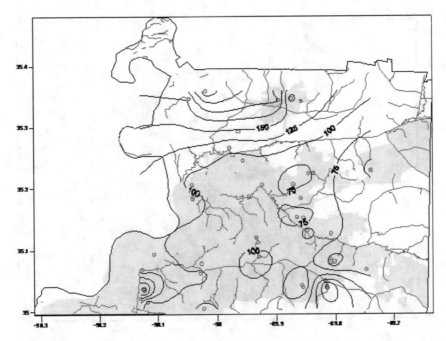

Fig. 6.6 Map showing TDS (mg/l) distribution, shallow aquifer, Memphis area (circles are well located, and gray zone is residential territory)

atmospheric precipitations, which are the main transporter of pollutants from the land surface to the water table.

Geochemical anomalies for Moldova's shallow groundwater are regional. Areas with TDS more than 1000 mg/l are not suitable for water use (Fig. 6.8). According to Moldova's standards, such areas need to be contoured with the isoline 1500 mg/l. In this case, areas not suitable for groundwater use significantly increase. Territorially, these areas are low topographical places, where the water table is not far from the land surface (usually until 3.0 m). Consequently, such places are characterized by a pollution vulnerability risk of medium and high. Analogous to the Memphis area, Ca is regionally distributed (Fig. 6.8). High Ca concentrations are associated with anomalous values of TDS.

Rastatt area is well hydrogeochemically studied (Wolf et al. 2007). The map of water electrical conductivity is functionally correlated with TDS. Investigated areas have the selective vulnerability risk. High values of TDS are associated with medium and high vulnerability risk for an unconfined aquifer. For this area, the urban sewage system is the main source of pollution. Nevertheless, the town itself is also one of the strongest pollutants of point and diffusive sources. Medium risk of

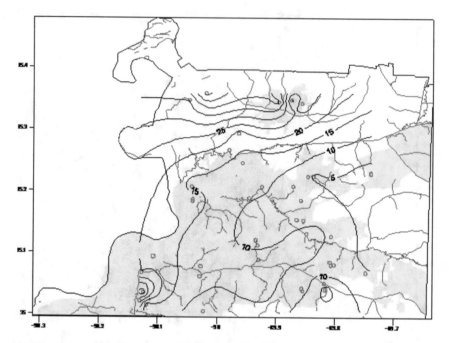

Fig. 6.7 Map showing Ca (mg/l) distribution, shallow aquifer, Memphis area (circles are well located, and gray zone is residential territory)

groundwater pollution is reflected in the high values of TDS (as EC) and Cl (see Figs. 5.20, 5.21, Chap. 5). Additionally, the map of nitrate concentrations is a direct indicator of the vulnerability risk and pollution (Fig. 6.9).

Figures 6.3, 6.4, and 6.5 are nomographs of the Moraru classification constructed using the minimum and maximum statistical values of major chemical elements and TDS. The dark areas of the nomographs indicate the range of possible concentrations of major elements as a percent of TDS. Detailed examination of nomographs suggests the following: The chemical composition of shallow groundwater in the Memphis area is very changeable. The TDS is in the interval of about 50–1000 mg/l, and the average water type is F1-F3NaHCO$_3$. There is also F1-F3NaCl type in limited quantities as well.

The groundwater of Moldova differs significantly from that of the Memphis area. Moldova's shallow water includes many types under the Moraru classification, namely F1-B3 Na(Ca)HCO$_3$. Variations of TDS from about 60 to 5000 mg/l suggest that water types also vary. The subtypes F1-F3 are primarily NaHCO$_3$, and subtypes F4-B3 are predominantly Ca(Na)SO$_4$.

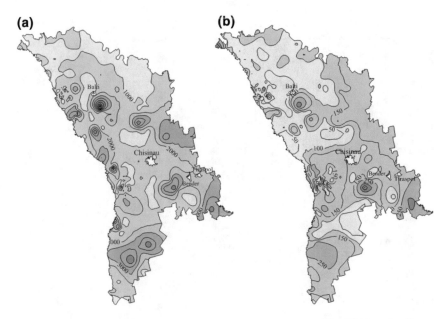

Fig. 6.8 Maps showing TDS (**a**), and Ca (**b**) distribution shallow aquifer, Moldova (mg/l)

The shallow aquifers of Rastatt area have simple hydrogeochemical conditions. TDS is varying from 90 to 600 mg/l. Water type is as F1-F3 CaHCO$_3$. This indicates that process of pollution is only in the first stage. The most variable is HCO$_3$. This element is geochemically driven by atmospheric precipitation, the pH-Eh system, and by the presence of CO$_2$ in the aquifer. Wolf et al. (2007) note that the distribution of the CO$_2$ partial pressure is normal in the probability net and the pCO$_2$ medians of the three data records are between -1.74 and -1.82. In the probability net, the distribution of the selected 118 groundwater analyses shows a sharp bend at a log pCO$_2$ of about -1.6. This can be interpreted as a boundary between CO$_2$-rich groundwater and groundwater without aggressive carbonic acid. These authors continue, "in the probability net the distribution of the selected 118 groundwater analyses shows sharp bends at bicarbonate contents of about 135 and 388 mg/l. The beginning of the unambiguously anthropogenically influenced area starts with hydrogen carbonate concentrations higher than 388 mg/l in the Rastatt model area."

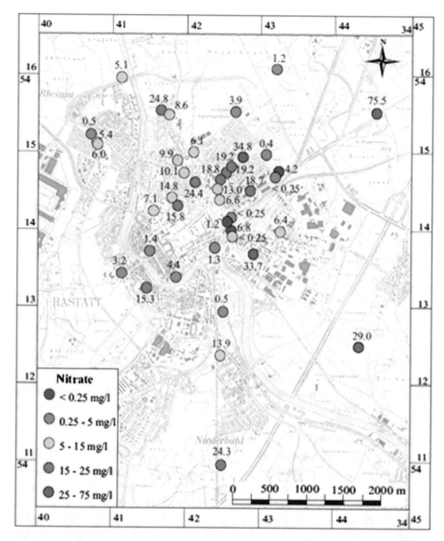

Fig. 6.9 Maps showing nitrate distribution, shallow aquifer, Rastatt, Germany (mg/l) (Wolf et al. 2007)

References

Aitchison, J. (1981). A new approach to null correlations of proportion. *Mathematical Geology, 13* (2), 175–189.

Aitchison, J. (1986). *The statistical analysis of compositional data* (p 380). Bristol: J. W. Arrowsmith Ltd: .

Chebotarev, I. I. (1955). Metamorphism of natural waters in the crust of weathering. *Geochimica and Cosmochimica Acta, 8,* 22–48.

Hem, J. D. (1985). *Study and interpretation of the chemical characteristics of natural water* (3rd ed., Vol. 2254, p. 264). U. S. Geological Survey Water Supply Paper.

Moraru, C. E. & Anderson, J. A. (2005). A comparative assessment of the ground water quality of the Republic of Moldova and the Memphis, TN Area of the United States of America. Ground Water Institute: Memphis, TN, p 188.

Pitiova, C. E. (1978). Gidrogeohimia (p. 210). Moscow: MGU.

Pitiova, C. E. (1984). Gidrogeohimiceskie Aspecty Ohrany Geologiceskoi Sredy (p. 190). Moscow: Nedra.

Posohov, E. V. (1975). Obshaia Gidrohimia (p. 140). Leningrad: Nedra.

Samarina, V. S. (1977). Gidrogeohimia (p. 280). Leningrad: LGU.

Valiashko, M. G. (1939). Contribution to our knowledge of the main physicochemical mechanism governing the development of Saline Lakes. (Vol. 23, No 7). Doklady AN SSSR.

WHO (World Health Organization). (2011). *Current edition of the WHO guidelines for drinking-water quality* (4th edn.). http://www.who.int/water-sanitation-health/dwq/guidelines2/en/.

Wolf, L., Klinger, J., Hoetzl, H., & Mohrlok, U. (2007). Quantifying Mass fluxes from urban drainage systems to the urban soil-aquifer system. *Journal of Soils and Sediments, 7*(2), 85–95.